TECHNICAL WRITING

Structure, Standards, and Style

ROBERT W. BLY

GARY BLAKE

McGraw-Hill Book Company

New York   St. Louis   San Francisco   Bogotá   Guatemala
Hamburg   Lisbon   Madrid   Mexico   Montreal   Panama
Paris   San Juan   São Paulo   Tokyo   Toronto

1 2 3 4 5 6 7 8 9 0 F G F G 8 7 6 5 4 3 2

ISBN 0-07-006174-2 {H.C.}
ISBN 0-07-006173-4 {PBK.}

LIBRARY OF CONGRESS CATALOGING IN PUBLICATION DATA

Bly, Robert W.
  Technical writing.
  1. Technical writing. I. Blake, Gary.
II. Title
T11.B63   808'.0666   82-15223
ISBN 0-07-006174-2   AACR2
ISBN 0-070006173-4 (pbk.)

Book design by Victoria Wong

For my parents, my sister Fern, and for Amy

—RWB

For Jack Nathan

—GB

Contents

CHAPTER 7: Producing the Technical Document   91

APPENDIXES

Acknowledgments

In a sense, every book is a collaboration. We would like to acknowledge the help of several silent "collaborators" who lent their support and efforts to this project. Special thanks to Jack Nathan for checking the manuscript and providing dozens of helpful comments and suggestions. Thanks also to Eve Blake, who made numerous helpful comments on each draft of this book, and to Dr. Stephen Aaron for his guidance.

We are indebted to our agent, Dominick Abel, for his encouragement, patience, and support; and to our editor, Tim Yohn, who shared our vision of this book and of its potential.

Introduction

When we first conceived of *Technical Writing: Structure, Standards, and Styles*, we did what all would-be authors do: we looked to see what had been written on the subject. To our amazement, about 100 books were in print on some facet of technical writing.

It seemed almost foolish, in the face of such competition, to continue with the project. But we did, believing that the ever-growing numbers of engineers, scientists, sociologists, computer programmers, ecologists, psychologists, and other technically oriented writers still lacked a single, authoritative source of accurate, up-to-date information. We wanted to write a brief, easy-to-read style guide which would be accessible to both technical and nontechnical writers.

We also felt strongly that the book should combine the breezy flow of popular nonfiction with the scholarly foundation needed to appeal to college students and the scientific community.

Technical Writing: Structure, Standards, and Styles looks at writing through the perspective of a technical writer. The rules are illustrated by examples drawn from technical literature in a wide variety of sciences. This book addresses problems that come up every day in the work of technical writers.

Chapter 1 provides a checklist of seven qualities which an effective technical document should have. Some of these also apply to nontechnical writing, but examples and essays explain the special relationship of these qualities to the technical composition.

Chapter 2 discusses in detail the whole range of technical writing tasks, from straight technical research reports to trade journal advertisements and sales brochures.

Chapter 3 covers the proper use of numbers, units, equations, and symbols. It will help keep mathematics correct and consistent in your writing.

Chapter 4 covers the basics of grammar—but we emphasize rules that will help you avoid the most common errors in technical writing.

Chapter 5 presents the basic principles of technical composition. It discusses the tone, style, and voice used in the various types of technical writing.

Chapter 6 shows you how to avoid jargon, clichés, antiquated phrases, and other evils.

Finally, Chapter 7 tells the technical writer how to take a manuscript through typing, editing, revisions, approvals, design, and layout to final printing and publication. It will help novices get through this ordeal and give seasoned pros a few tips they may not know.

Appendix C discusses how technical writers find work—an important topic that, we believe, has not been addressed in any other work on technical writing.

We look forward to an era of unprecedented growth in the sciences. The advances in genetics, cable television, computer technology, medicine, energy sources, and ecology over the past 10 years have been astounding, and the next 20 years promise even greater growth, hundreds of thousands of new jobs in the sciences, and an increased emphasis on clear, concise, accurate scientific writing.

We hope that you find *Technical Writing* enjoyable as well as handy. We tried to follow our own precepts, keeping the book brief, lively, and to the point. We combed through numerous style guides and reference books in search of definitive rules, hints which show how the language has changed and is changing, and suggestions as to how writers can eliminate poor stylistic habits.

Many other thoughts are in our minds as we write this introduction: the perils of writing about writing, our hope that we have covered *your own* technical field in the examples, our concern that we have steered a fair course between being authoritative and open-minded. But, for the sake of brevity, we'll stop here.

Elements of Technical Writing

> Newspaper reporters and technical writers are trained to reveal almost nothing about themselves in their writing. This makes them freaks in the world of writers, since almost all of the other ink-stained wretches in that world reveal a lot about themselves to the readers.
>
> — KURT VONNEGUT, Jr., novelist and former technical publicist for General Electric

Most technical writers would hardly classify themselves as "freaks in the world of writers." Yet technical writing—the literature of science and technology—is different from journalism, fiction, advertising copywriting, and other nontechnical prose.

Just what is technical writing? To begin, technical writing is defined by its subject matter—it is writing that deals with subjects of a technical nature. By *technical* we mean anything that has to do with the specialized areas of science and industry.

Traditionally, technical writers are thought of as "engineering writers." But in addition to engineering and applied sciences, technical writers are involved in all areas of physical, natural, and social sciences including anthropology, archaeology, biology, botany, earth science, ecology, geology, management science, medicine, psychology, sociology, and zoology, to name a few.

Because technical writing usually deals with an object, a process, or an abstract idea, the language is utilitarian, stressing accuracy rather than style. The tone is objective; the technical content, not the author's voice, is the focal point.

The difference between technical writing and ordinary composition is more than just content, however. The two differ in purpose as well. The primary goal of any technical communication is to *accurately transmit technical information*. Thus it differs from popular nonfiction, in which the writing is intended to entertain, or from advertising copywriting, which exists to persuade. Technical writers are concerned with communication; and if they have to, they will sacrifice style, grace, and technique for clarity, precision, and organization.

There are no set guidelines that clearly define the difference between technical writing and nontechnical writing. But we can offer a checklist of seven characteristics common to all good technical writing. If your technical writing doesn't have all these qualities, it probably will not serve its intended purpose as well as it should.

GOOD TECHNICAL WRITING IS . . .

1. Technically accurate

Since the intended goal of any technical writing is the transmittal of technical information, even the most well-written technical document is ineffective if the facts, theories, and observations presented are in error. The content must be true and as scientifically accurate as is humanly possible. Technical writing that contains technically inaccurate statements reflects inadequate knowledge of the subject and poor use of language.

Why is technical accuracy more important in technical writing than in popular magazine articles, books, and other nonfiction? Technical documents are not merely leisure reading; their readers make business decisions and scientific judgments based on the data presented.

An error in a Sunday supplement newspaper story may result in misinformed readers, and nothing more. Errors in technical documents can cost industry hundreds of thousands of dollars, and the results of good scientific work can be obscured by hastily prepared reports that are full of inaccuracies.

2. Consistent

"A foolish consistency," wrote Ralph Waldo Emerson, "is the hobgoblin of little minds." Maybe so. But inconsistencies in technical writing will confuse readers and convince them that your scientific work is as sloppy and unprofessional as your prose.

Today technical literature is plagued with random and unnecessary capitalization, mixed sets of units of measure, and indiscriminate use of abbreviations, punctuation, and rules of grammar. Consider this example:

> These mist eliminators are available in diameters as large as 7 to 8 feet in diameter, but are usually limited to 36–42″ diameter in sugar refinery applications. The mesh pads are usually 12 to sixteen cm thick. In most sugar refinery installations, they will eliminate B.O.D. and cut product loss up to ninety-six percent. As you know, BOD is caused by decomposing organic matter in wastewater streams.

Why, asks the puzzled reader, are diameters given in English units (feet and inches) while thickness is given in metric units (centimeters)? Why write *B.O.D.* in one sentence and *BOD* in the next? Why write out *sixteen* when *12* is written in numerals?

It takes a careful editor to keep everything in one consistent style. The reward of uniformity is a technical document that reads as if it were written by an educated, literate person.

3. Clear

In fiction, essays, and other nontechnical writing, clarity is not necessarily the key to good style. Obscurity has its place in some forms of literature, but obscurity has *no* place in technical literature, where comprehension is the chief concern.

Technical writers are successful if their work can be readily understood by the intended audience. Unfortunately, most technical writers never stop to consider who this audience is, and they write as if all their readers had the same level of technical expertise as they. The result is complicated, difficult-to-understand technical papers and reports that are full of jargon and run-on sentences and are poorly organized.

How can you write to be understood? Here are a few simple rules:

Keep the writing short and simple. Use small words, not big ones. Keep paragraphs and sentences short. Break up the writing into short sections, and try to limit each section to one central theme or idea. (Sentences, paragraphs, and compositions that express one idea are easier to understand than those that express multiple ideas.)

Avoid jargon. Some technical terminology is valid and necessary. But too much technical jargon makes writing incomprehensible. Remember, not every reader knows all the latest "buzzwords" in a specialized field. And your readers are busy people, more likely to put your report aside than to reach for a technical dictionary.

Know your readers. Typically, a single technical document is distributed to readers of widely varied levels of technical expertise—research scientists, applications engineers, sales representatives, purchasing agents, top-level executives, consultants. Write so that the most nontechnical reader can understand you easily. Do not think that this will offend technically oriented readers; we have never heard a scientist or an engineer complain of a report or paper that was too easy to read.

4. Correct in spelling, punctuation, and grammar

All writing—with the possible exception of dialect, poetry, and experimental fiction—must follow the standard rules of spelling, punctuation, and grammar laid down in the *U.S. Government Printing Office (GPO) Style Manual, The New York Times Manual of Style and Usage,* and other standard references on the English language.

Many scientists and engineers are not overly concerned with these seemingly picayune matters of English usage. After all, they reason, science and technology—not periods, parentheses, and participles—are of interest to technical people.

Unfortunately, even the most indifferent technical writers are quick to spot misspelled words and sloppy grammar. These errors point to authors who are either lazy or uncaring about their work. And that is simply not acceptable in today's highly professional business and academic environments.

5. Concise

Like most people in industry and academia, your readers are extremely busy. Concise technical writing is easier and less time-consuming to read than wordy technical writing.

By avoiding undesirable repetition, wordy phrases, overly general words, and jargon, you can produce succinct, readable prose without sacrificing technical content. Your words and phrases should be precise, your composition well organized and to the point. The example in the left column lacks conciseness and should be rewritten as shown in the right column.

It is also essential that the interior wall surface of the conduit be maintained in a wet condition, and that means be provided for wetting continually the peripheral interior wall surface during operation of the device, in order to avoid accumulation of particulate matter about the peripheral interior surface area.

The interior wall must be continuously wetted to avoid solids buildup.

Many technical writers think that long, jargon-filled reports seem more substantial and important than simple, concise writing. They are wrong. The length of a piece of writing is no guide to its importance. The First Amendment to the U.S. Constitution contains 45 words, and Newton's first law of motion contains 29. In 1976, Jacqueline Jones of Lindale, Texas, wrote a letter to her sister, Mrs. Jean Stewart, of Springfield, Maine. It was 1,113,747 words long.[1]

---

[1] Norris McWhirter, *Guinness Book of World Records: 1979 Edition,* Sterling, New York, 1979, p. 217.

6. Persuasive

The primary goal of technical communication is to transmit information; its secondary goal is to persuade.

All technical documents have a selling job to do: Technical papers and articles promote a company's technical expertise and capabilities or support a university's request for grants. Product information sells industrial equipment. Technical reports help scientists and engineers sell their projects to company management. The quality of a technical proposal can influence the outcome of a contract award.

Still, the product, process, or idea—not the author—is the focus of the technical document. The most persuasive technical documents offer a clear, objective presentation of the relevant facts that support your ideas, opinions, and recommendations.

People in technical fields tend to make decisions based more on logic and rational thought than on emotion; cuteness, puffery, boasting, and other high-pressure selling tactics do not work well in the technical marketplace.

7. Interesting

"It is a sin to bore your fellow creatures," claims David Ogilvy, founder of Ogilvy & Mather, one of the world's largest advertising agencies. Ogilvy knows that business documents must gain and keep the reader's interest if they are to be read. Your technical report competes with many other communications, such as letters, memorandums, trade journals, popular magazines, newspapers, and novels. Be lively and lucid, not dull and boring. People in technical fields are human, too.

2

Tasks of the Technical Writer

In most organizations, there are two kinds of technical writers. The first includes managers, executives, engineers, and scientists who have to spend some part of their day writing technical documents. Not writers by profession, these people often view writing as an unpleasant, time-consuming chore—something that takes them away from their *real* work at their desk or laboratory bench.

The second kind includes full-time professional writers and editors who work in the organization's technical publications department or reports and proposals group. It's their job to write, edit, and prepare technical documents for publication.

Both the professional technical writer and the business executive get involved in a wide variety of technical writing tasks, including letters, manuals, reports, abstracts, proposals, articles and papers, presentations, product literature, and advertising. Although the same guidelines for good technical writing apply to all these documents, each is unique in its audience, purpose, technical depth, format, and style. That's why this chapter has separate discussions of each type of publication.

LETTERS

A recent TV commercial informed viewers that the U.S. Post Office handles 300 million pieces of mail *every day*.

That's a lot of letters.

And letters are an important form of communication in the business and scientific communities, where no assignment is complete unless it's put in writing and transmitted to the right people.

Here are a few of the different kinds of letters technical professionals have to write:

> Letters of technical information
>
> Letters of transmittal for reports and proposals
>
> Letters of instruction
>
> Announcements of new products, facilities, policies, and services
>
> Answers to product inquiries

Years ago, letters were written in stiff, formal language. Now business communications are less stuffy and more personal. Businesspeople recognize that a letter is a personal communication from one human being to another . . . not from one inanimate, corporate entity to the next. Effective business letters get their messages across by being friendly and helpful.

Still, many executives—especially those with technical backgrounds—hastily dictate letters in the pompous "corporatese" style. Do not treat your letters as doctoral dissertations. Instead, write as if you were conversing with the person who is going to receive your correspondence.

Writing in an easygoing, conversational tone takes practice. But until this style comes naturally, you can make your letters easier to read by avoiding these letter writer's clichés:

> *Hackneyed Expressions in Business Letters*
>
> Enclosed please find
>
> Attached please find
>
> I am forwarding herewith
>
> We are in receipt of your letter dated
>
> Per your inquiry of
>
> In reply to your letter of (date), in which you stated that
>
> If you will kindly inform us
>
> In accordance with your wishes
>
> Awaiting your earliest reply, we remain

> In view of the above
> You will please note
> Please don't hesitate to call
> In view of the foregoing

Let's take a look at two letters that show the difference between old-fashioned corporatese and today's more personal approach. Both are answers to an inquiry about the company's product.

SPARTAN CO., INC.
770 LEXINGTON AVENUE
NEW YORK, N. Y. 10021

October 23, 1980

Mr. Ron Brick
Chief Engineer
Chemtech Corp.
130 Sumner Ave.
Akron, Ohio 44309

Dear Mr. Brick:

CHEMICAL PROCESSING magazine has informed us of your interest in the level detectors manufactured by our firm for use in the chemical processing industry. As you may perhaps know, we are one of the oldest and most well-respected manufacturers of such equipment, and our product line includes the following types of level detectors: Beam Breaker, Bubble, Diaphragm, Capacitance, Conductive, Differential Pressure, Displacer, Float and Float and Tape, Glass and Magnetic Gauge, Hydrostatic Pressure, Inductive, Infrared, Microwave, Optic Sensor, Paddle, Pressure-Sensitive, R-F Admittance, Radiation, Sonic Echo, Strain Gauge, Thermal, Tilt, Vibration, and Weight and Cable level detectors that are described in the enclosed technical sales literature.

Since you may also have requirements for our other types of process equipment, we are enclosing our All-Line Catalog and Data Sheets with the request that you fill in the Data Sheets with as much information as you have availble, returning them to us for the consideration and recommendations of our Engineering Department, enabling us to quote you , if possible, on specific applications. Finally, as our Company is now in its fourth decade of continuous service to its many Customers in this Country and Abroad, we are sending along a reprint of our latest annual report which will give you more information on our activities. We will await with interest your specific inquiries. Thank you once again for contacting us.

Very Truly Yours,

John N. Guterl

John N. Guterl, President

Mars Mineral Corp.

P.O. BOX 128 • VALENCIA, PENNSYLVANIA 16059
412-898-1551 • TELEX 866452

September 18, 1981

Mr. L. Moore, Proj. Engr.
SPAUTAN CO.
771 Lexington Ave.
New York, NY 10021

Subject: Pelletizing Information

Dear Mr. Moore,

Thanks for your interest in our Pelletizers. Literature is
enclosed which will give you a pretty good idea of the simplicity
of our equipment and the rugged, trouble-free construction.

The key question, of course, is the cost for equipment to
handle the volume required at your plant. Since the capacity of our
Pelletizers will vary slightly with the particulates involved, we'll
be glad to take a look at a random 5 gallon sample of your material.
We'll evaluate it and get back to you with our equipment recommendation.
If you will note with your sample the size pellets you prefer and the
volume you wish to handle, we can give you an estimate of the cost
involved.

From this point on we can do an exploratory pelletizing test,
a full day's test run or we will rent you a production machine with
an option to purchase. You can see for yourself how efficiently it
works and how easy it is to use. Of course the equipment can be
purchased outright too.

Thanks again for your interest. We'll be happy to answer any
questions for you. Simply phone or write.

Very truly yours,

MARS MINERAL CORPORATION

Robert G. Hinkle
Vice President, Sales

The letter from the Spartan Company is fairly straightforward and comprehensible. But would you ever talk to a customer in such long-winded sentences? (The second sentence in the first paragraph is 82 words long.) The letter is full of letter writer's clichés (. . . *has informed us of your interest* . . ., *We will await with interest your specific inquiries*). Words such as *company, country,* and *abroad* are capitalized for no reason. Plus, this letter has "form letter" written all over it.

Mars Mineral Corp. does much better with its livelier, more lucid reply letter. The letter is friendlier, the paragraphs and sentences are shorter, and the tone is more conversational—like one friend talking to another. While the letter from the Spartan Company merely repeats information found in the bulletins mailed with the letter, Mars Mineral suggests a course of action (sending in a material sample for evaluation) that can solve the customer's problem and lead to the sale of the Mars Mineral pelletizer.

MANUALS

Many large engineering firms have departments that produce technical manuals about the company's products. *Instruction manuals* tell how to install, maintain, and operate equipment. *Theory-of-operation manuals* describe the principles of the operation in greater detail. They may include schematic diagrams, blueprints, equations, tables of operating data, performance curves, and anything else that tells an engineer how and why a system, product, or process works. *Sales manuals* provide technical salespeople with the product specifications, pricing, and other information needed to sell products.

Like a well-written cookbook, these manuals present instructions for completing a certain task. The only difference is that the operations described in technical manuals are usually highly complex.

Still, the technical writer must strive to make manuals clear, direct, and as easy-to-follow as, say, this cookbook recipe for chicken soup.

In a large kettle, bring 6 quarts of water to boil. Add a cleaned chicken and simmer for 30 to 60 minutes. Then add vegetables, cover the kettle, and simmer for another 60 minutes. When the soup is done, remove the chicken and all vegetables except the carrots. Serve with rice or noodles.[1]

A dissection manual for high school science students approaches the complex operation of dissecting a frog in the same simple style

[1] *Our Favorite Recipes,* Women's American ORT, New Jersey District III, 1974, p. 28.

as the chicken soup recipe. The writer didn't make the instructions needlessly complicated or full of jargon simply because the subject is technical.

> The heart is encased in a thin sac called the *pericardial sac*. Cut through the thin membrane of this sac with the tip of a very sharp razor. Do not cut the heart itself. Then spread the membrane with forceps to expose the heart.[2]

As you can see, instruction manuals are written in the imperative mode. Be direct. It's better to write *Connect the communications line* than the weaker form *The communications line should be connected*. Effective instructions, as you can see in the sample below, *tell the reader what to do in the simplest, most direct language possible*.

> Connect one end of the line cord with J3 on the back panel. This cord includes a ground wire for plugging into a grounded outlet. Before applying power, make sure the power supply in the equipment conforms to the form of primary power available. Required fuses are listed below.

If the operation is clearly a step-by-step procedure, you can make life easier for the reader by writing the instruction manual as a series of numbered steps.

Startup Operation for Polymer Mixer

1. With the injection unit fully retracted, bring the barrel and mixer to operating temperature.
2. Set the machine operation mode to manual and the boost and secondary pressure regulators to their lowest settings.
3. Quickly depress and release the injection switch. If the screw bounces back, allow more heat soaking time.
4. Once the polymer flows freely in a purging mode, increase the injection pressure as required and begin the molding operation.

Thus, if operators need to discuss the procedure, they can ask a question by referring to step 3 instead of "the fifth line down in the second paragraph on the fourth page, where it says to release the whatchamacallit switch."

[2] William Berman, *How to Dissect: Exploring with Probe and Scalpel*, Sentinel, New York, 1961, p. 105.

Here's a surefire method of ensuring that your manual communicates effectively: Give a copy to an operator and see whether she or he can install or operate the equipment by reading your instructions. If the operator becomes hopelessly lost, either your writing is unclear or your instructions are incorrect. If the operator gets things running smoothly without too much trouble, you've done your job well.

PROPOSALS

A *technical proposal* is a document prepared by an organization to sell its services, products, or ideas. In the proposal, the organization offers to provide services, products, or ideas to the potential buyer within a certain time and at a specified cost.

Both the U.S. government and many corporations solicit proposals from bidders before awarding a contract. These contracts may bring in millions of dollars of work—projects that will command the resources of the company for months, and sometimes years.

The proposal is often the most important factor bearing on the winning of a contract. The survival of any organization, then, depends on its ability to prepare proposals that sell.

When the government and some corporations invite bidders to submit proposals for a project, they issue a document called the *request for proposal* (RFP). The RFP outlines in great detail the points that must be addressed in the proposal. (They can even go so far as to tell you how to type the proposal!) Technical writers use the RFP as an outline for the topics to be covered in their technical proposals.

For large contracts, the final technical proposal can be a multivolume document thousands of pages long. Naturally, large proposals are not written by a single individual; they are put together by a proposal *team*.

The professional technical writer usually heads the team as proposal manager. It's the manager's job to coordinate the writing and production of the document. The authors are the many engineers and technical managers who will produce the proposed system if the contract is won. Each author writes the sections of the proposal dealing with his or her area of technical expertise. A large proposal can have dozens of contributing authors; it's the proposal manager's job to mold their work into one clear, coherent text.

Often, the final document is published as two separate proposals:

1. A *technical proposal* presents the engineering solution and technical specifications for the customer's problem independent of the cost.
2. A *contractual and cost proposal* covers the total project cost (broken down by phases or steps). Costs can include materials, labor, research and development, and administration.

Sometimes, a third volume is added:

3. The *executive summary* highlights the key points in a short, nontechnical document for the busy high-level executive who lacks the time to read the whole proposal.

We said that you must produce effective proposals to win contracts and stay in business. The successful proposal is thorough, technically accurate, and responsive to the customer's requirements as specified in the RFP. It must be coherent, well organized, and written so that the key points can be picked out quickly and easily by proposal evaluators as they scan the pages.

Proposal manager Charles C. Anderson[3] offers these tips for proposal writers:

Before you start . . .

1. Treat the proposal as a sales document.
2. Understand the selling strategy.
3. Understand your assignment.
4. Study the RFP.

As you write . . .

5. Follow your writing plan.
6. Incorporate references.
7. Develop your topic logically.
8. Use summaries and visuals to reinforce your text.
9. Write clearly and concisely.

[3]Charles C. Anderson, "Twelve Tips for Proposal Writers," *EDN*, November 15, 1970, pp. 53–56.

After writing . . .

10. Review the text and artwork thoroughly.
11. Compare your draft and the published version.
12. Keep a proposal file.

By following Anderson's tips and the rules for good technical writing presented in this book, you can produce a proposal that gets your message across and brings in contract awards.

TECHNICAL ARTICLES AND PAPERS

There are more than 6,000 business, technical, academic, and trade publications in the United States. They publish hundreds of thousands of technical articles and papers each year.

Technical publications are the medium through which engineers and scientists tell their peers in other organizations about their work. In academic journals, such as the *Bulletin of the Atomic Scientist* or the *Journal of Applied Physics,* the subject matter is usually theoretical, and the authors and their readers are scientists doing research. Trade journals, however, prefer a more practical approach. The articles in *Chemical Engineering, Machine Design, Elastomerics,* and other trade publications provide information that helps engineers and managers in industry do their jobs better.

Technical articles, then, are written *by* technical professionals *for* technical professionals. Why would a busy research scientist or plant engineer take the time and trouble to write for publication? We can think of at least six reasons:

1. Personal satisfaction.
2. Increase in the author's status as a technical expert.
3. Good publicity for the author's company.
4. Professional prestige to increase the author's chances for promotion.
5. Writing teaches the author more about the subject.
6. The author is contributing to the pool of technical knowledge and helping others to learn.

The technical professional is not expected to write with the flair and style of a professional writer. Journal editors will polish and rewrite the engineer's manuscript to make it more interesting and readable.

But there's no point in writing for publication unless your piece is going to be read. To get the readers' attention, it's a good idea to "hook" them with a strong opening paragraph. Then make sure you *keep* their attention—by piling one fascinating fact on top of the next, as in this article from *Design News*.

> ### Lightwave Communications
> by F. W. Tortolano
>
> By the end of this year, installations planned by the Bell System will prove that communication by lightwave—sending information in bursts of light over hair-thin fibers—no longer is science-fiction fantasy. It is technology ready for widespread service.
>
> The new installations include a 40.6-mile system between Pittsburgh and Greensburg, PA, capable of carrying 25,000 simultaneous conversations, several times the call-capacity of more traditional copper cable. Other installations scheduled for 1981 include . . . . [4]

Surely, any technical manager with a professional interest in fiber optics would be tempted to read further.

Many corporations, including Raytheon, International Paper, and Westinghouse, publish their own magazines for distribution to customers and employees. The articles in these "house organs" are shorter, less technical, and "newsier" because many of the readers are laypeople. Here is an excerpt from an article published in the *Westinghouse Defense News* magazine:

> ### Laser Being Developed for Underseas Use
>
> Under a contract to the Office of Naval Research, Westinghouse is conducting research to determine the feasibility of a visible blue-green laser for underseas applications.
>
> To achieve such a laser, researchers indicate, would require the use of the metal excimer family as the laser source. In particular, researchers are looking at the thallium-mercury (TlHg) molecule to achieve their goal.
>
> According to the project manager, Dr. Bud Weaver of the Westing-

[4]F. W. Tortolano, "Lightwave Communications II," *Design News,* July 6, 1981, p. 72.

house Research and Development Center in Pittsburgh, excimers are unlike ordinary molecules because they are stable in their excited state, rather than in the ground state. This makes the excimer a nearly perfect laser molecule.

A unique characteristic of the developing TlHg laser is its "tuneability"—the ability to continuously vary the wavelength of laser light over the blue portion of the spectrum.[5]

House organs usually generate stories from within the corporation. But trade publications are always on the lookout for new and exciting stories from industry and academia. These can include case histories, industry roundups, market trends, controversial issues, new products, improved technologies, ongoing research and development, new manufacturing techniques, energy-saving ideas and systems, and service, engineering, or performance stories. If you would like to see your name in print, contact the editor of the journal for which you want to write and ask how to go about submitting an article. Although you won't gain the fame and fortune that best-selling novelists do, technical publications have their own reward.

REPORTS

Technical reports are the documents in which engineers, scientists, and managers transmit the results of their research, fieldwork, and other activities to people in their organization. Here's what the University of Rochester's department of chemical engineering has to say about engineers and report writing:

The importance of being able to write a good report cannot be emphasized too strongly. The chemical engineer who carries out an investigation or study has not completed his job until he has submitted a report on the project. The true value of the project and the abilities of the investigator may be distorted or unrecognized unless the engineer is able to furnish a commendable report.[6]

[5]"Laser Being Developed for Underseas Use," *Westinghouse Defense News*, September 1979, p. 12.

[6]"Guidelines for Technical Reports," University of Rochester, Department of Chemical Engineering.

Often, a written report is the only tangible product of hundreds of hours of work. Rightly or wrongly, the quality and worth of that work are judged by the quality of the written report—its clarity, organization, and content. Therefore, it pays to take the time to write a good report.

Professionals in business, government, and education produce several kinds of reports:

1. *Periodic reports* are submitted at regular intervals to provide information on the activities or status of the organization. Bank statements, annual reports, and call reports are examples of periodic reports.
2. *Progress reports* are updates on an ongoing activity as it is being accomplished. The activity may be construction, expansion, research and development, production, or other projects.
3. *Research reports* present the results of the research, studies, and experiments conducted in the laboratory.
4. *Field reports* present the results of an on-site inspection of some field activity, which might be construction, pilot-plant tests, or equipment installation and startup.
5. *Recommendation reports* are submitted to management as the basis for decisions or actions. They make recommendations on subjects such as whether to fund a research program, whether to hire more process engineers, whether to acquire a company.

Although reports can take many forms, most contain the following major sections: cover and title page, abstract, table of contents, summary, introduction, body (theory, apparatus and procedure, results, and discussions), conclusions and recommendations, nomenclature, references, and appendixes.

Let's take a brief look at each of these sections.

1. Cover and title page

The cover and title page are the reader's first impression of your report. The cover should be attractive, but not gaudy. The title page should be neat; the title should tell the reader exactly what the report is about.

2. Abstract

An abstract is a one-paragraph statement of the contents of the report. Abstracts are so important that we devote the next section of this chapter to them.

3. Table of contents

The table of contents lists every section heading and subheading and the page number on which it appears. Tables, figures, charts, and graphs are listed separately at the end of the regular table of contents.

4. Summary

While the abstract gives the reader enough information to decide whether to read the report, the summary presents its entire contents in a few hundred words. It covers what was done, how it was done, and all the important results.

5. Introduction

The introduction tells readers—including those not familiar with the subject matter or the reason for writing the report—the *purpose* of the report. It provides background material, theory, and explanation of why the work was done and what it accomplished.

6. Body

The body of the report contains the detailed *theory* behind the work; outlines the *apparatus* used and the *procedures* followed in the experiments; presents experimental *results,* data, and observations; and provides a *discussion* of the meaning, significance, accuracy, and application of these results.

7. Conclusions and recommendations

The *conclusions* are a series of numbered statements which show how the results answered questions raised in the stated purpose of the research. On the basis of the results and conclusions, the writer can make *recommendations* as to the need for further research or commercial application of the work.

8. Nomenclature

This section lists, in alphabetical order, the symbols used in the report. Units of measure for each symbol are presented, too.

9. References

An alphabetical bibliography shows all the technical literature (books, papers, brochures, and reports) used by the author in the work and report.

10. Appendixes

Appendixes contain any sample calculations, tables, mathematical derivations, or calibration data that are too long, cumbersome, or unimportant to be included in the main body of the report.

These sections are standard, but their order may vary from company to company. (At General Motors, for example, conclusions appear at the front of the report, right after a brief foreword.) If you have to write a report, check with your supervisor or technical publications department to see whether your organization has guidelines for technical publications.

ABSTRACTS

At a recent meeting of the American Institute of Chemical Engineers (AIChE), more than 316 technical papers were presented on topics ranging from the "Exposure of Steel Containers to an External Fire" to "Thermodynamic Availability Analysis in the Synthesis of Energy-Optimum and Minimum-Cost Heat Exchanger Networks." Prior to the meeting, AIChE members were mailed a 94-page technical program containing abstracts of all the papers. By scanning abstracts, meeting participants could quickly decide which presentations to attend.

An abstract is a short (generally 250 words or less) statement of the contents of a report, a paper, or other document. The abstract introduces the subject matter, tells what was done, and presents selected results.

Your readers cannot possibly read all the reports that come across their desks, but studies show that executives usually *do* scan the abstracts to see whether the report is of interest. A well-written abstract is the best means of convincing the right people to read your report.

Abstracts of thousands of technical papers presented each year are permanently bound for future reference in volumes such as *Science Abstracts, Abstracts on Hygiene,* and *Applied Mechanical Reviews.* Many technical writers are employed full-time in editing and indexing abstracts for these publications. (Before turning to science fiction, novelist Arthur C. Clarke was an assistant editor of *Science Abstracts.*)

Abstracts describe complex scientific research in fewer words than are on this page. Therefore, they must be extremely concise.

Here is a well-written abstract from a paper presented at a conference of pulp and paper engineers. It tells the whole story in four fact-filled sentences.

Energy Requirements for Coated and Uncoated Papers

A study was made of the energy required to produce coated and uncoated papers. Energy use requirements were determined for each process from pulping through coating. Raw material energy needs were included in the study. Based on the total energy concept, it was determined that the production of coated papers requires 5 to 12 percent less energy than the production of uncoated papers of similar weight and brightness.[7]

The abstract clearly states the important result in quantitative terms: producing coated paper takes 5 to 12 percent less energy than producing uncoated paper. However, we should have been told *how* energy-use requirements were determined; a good ab-

---

[7]Copyright © 1981. TAPPI. Reprinted from G. M. Hein and R. L. Lower, "Energy Requirements for Coated and Uncoated Papers," Technical Association of the Pulp and Paper Industry, "Instructions for Paper Preparation and Presentation at TAPPI Meetings," Atlanta, pp. 19, with permission.

stract always says how the work was done. In the abstract below, the work done, techniques used, and limits of precision are described in one tightly written paragraph.

A Precise Optical Instrumentation Radar

The instrumentation tracker described provides real-time positioned data on high-speed cooperative targets with precisions of ± 1 m at ranges between 300 m and 10 km. Unambiguous range is determined by a precise digital FM-CW ranging technique at a rate of 15 per second. A target-mounted beacon and a narrow laser ranging beam permit measurement of target position to values much less than target dimension. Azimuth and elevation angles are read out by precision shaft angle encoders and recorded in binary form, along with range and time, on a magnetic tape or directly into a real-time computer.[8]

ORAL PRESENTATIONS

From time to time, you may have to make a technical presentation to colleagues, management, or customers. These oral reports inform, motivate, train, instruct, and sell. By giving a talk, you can communicate effectively and persuasively to large groups. Oral presentations often have more impact than a written communication.

What are some common oral presentations in the technical fields? Research scientists present technical papers at professional meetings. Executives lead panel discussions and chair committees. Salespeople give sales presentations to customers and prospects. Engineers explain their work to managers in technical briefings. Top management discusses the year's business at annual employee meetings.

Oral presentations are two-way communications between you and your audience, and speaking gives you two advantages over the printed page:

1. Speakers, unlike writers, can immediately gauge audience reaction and response to their message.

[8]T. C. Hutchison, A. A. Hagen, H. Laudon, and C. R. Miller, "A Precise Optical Instrumentation Radar," *IEEE Transactions on Aerospace & Electronic Systems,* vol. AES-2, no. 2, March 1966.

2. Speakers can use diction, phrasing, tone, inflection, gestures, and "body language" to give their presentations added impact and emphasis.

An oral presentation is an opportunity to forcefully present your message to a broad and receptive audience. Do not bore and alienate your audience by *reading* your speech, report, or technical paper. Instead, *talk* to them. Tell them what is interesting and important about your subject. Stress the high points. And, when possible, illustrate your presentation with visual aids.

Today, few business and technical presentations are given without some type of audiovisual (AV) aid. The most popular AV aid is probably the 35 mm slide, followed by overhead transparencies, flip charts, films, videotapes, filmstrips, and the old schoolroom standard, the chalkboard.

Why AV aids? Because some things are communicated better in pictures than in words. For example, no spoken text can ever describe what a zebra looks like as precisely as a photograph of the animal. And a blueprint can tell a civil engineer more about the structure of a bridge than sentences and paragraphs can. The different types of visuals and the messages they communicate are listed here:

Visual	This Visual Shows
photograph or drawing	what something looks like
map	where it is located
diagram	how it works or is organized
graph	how much there is (quantity)
pie chart	proportions and percentages
bar chart	comparisons among quantities
table	a body of data

Visuals should be used only if they enhance the meaning of the spoken text. If information or a concept can be verbalized adequately, there is no need to visualize it.

Oral presentations require planning to be effective. The content must be organized in a logical manner. While you should not read a prepared text word for word, notes or an outline can help keep you on the right track. And you should practice your presentation until it's smooth, clear, and natural.

Timing is important. Speakers at briefings or meetings often are held to a strictly enforced time limit. As a rule of thumb, the average speaker talks at a rate of 100 to 150 words per minute. At this rate, it should take you 10 to 15 minutes to get through a 1,000-word text. (That's about equal to four typewritten pages.)

PRODUCT INFORMATION

In 1980, purchasing agents and other buyers paid more than $1,500,000,000,000 for industrial products and services. Of these buyers, 90 percent[9] said they required some kind of printed product information before making a purchasing decision. This printed material includes sales brochures, technical bulletins, case histories, catalogs, product data sheets, line cards, and other technical literature. Each is described below.

Sales brochures

These are distributed to potential customers who request information about the company or its products. As a sales tool, the brochure must get and hold the readers' attention and then persuade them to take action. (This action might be the placement of an order, a request for further information, or contact with a company sales representative.) Brochures emphasize key product features and selling points, and they omit extraneous technical detail. Today's sophisticated sales brochure is often produced by professional graphic designers who turn mundane documents into attractive, attention-getting publications.

Technical bulletins

After reading the sales brochure, a technical professional may want more detailed information. Technical bulletins go into greater technical depth than brochures and can include such detailed in-

---

[9] According to promotional material from the Thomas Publishing Company, publishers of *Thomas Register,* a widely circulated industrial buyer's guide.

formation as performance curves, dimensions, materials of construction, theory of operation, sample calculations, and tables of operating data. Technical bulletins usually address a specific application or product feature, while sales brochures tell the whole story.

Case histories

A case history is a technology "success story." It tells how a problem was solved through the use of a product, system, or process. Case histories can be published as articles in trade journals, used as the basis for advertisements, or printed as product literature.

Catalogs

Catalogs are descriptive lists of a company's products. They may be printed in booklet form or appear as advertising inserts in such industrial buyer's guides as *Thomas Register, U.S. Industrial Directory,* or *MacRae's Blue Book.* Catalog descriptions should include product photographs or drawings, features, dimensions, weights, and prices, when possible.

Product data sheets

Product data sheets present straight technical data with little or no explanation or editorializing. A product data sheet on a new grade of carbon black, for example, might include information about chemical and physical properties: tensile strength, specific gravity, tint, particle size, ash and sulfur content, nitrogen surface area, and pore density.

Line cards

Line cards or product guides list a manufacturer's or distributor's complete line of products. Usually they are printed as one-page inserts for the purchasing agent's file of suppliers.

Other technical literature

These can include reprints of trade journal articles and technical papers, as well as bulletins on special processes and applications.

ADVERTISEMENTS

The newspapers, magazines, and other publications you read at leisure run paid advertisements for products that you, as a consumer, might buy—anything from cars, cigarettes, and candy bars to soft drinks and stereo systems.

The technical, trade, and business publications you read at work also run paid advertisements. In 1980, they carried 1.9 million pages of "business-to-business" advertising which sold systems, products, and ideas to industrial buyers.

There are several basic differences between consumer and industrial marketing.

First, consumer buying is usually a one-step process. If you see a soft drink advertisement and want the product, you can go to the store and buy it. Purchasing an industrial product, however, is a business decision that takes many steps. McGraw-Hill's Laboratory of Advertising Performance (LAP) has found that it takes six sales calls to close the average industrial sale. The trade journal reader's most frequent response to an industrial advertisement, says LAP, is not to buy, but to request more information . . . usually a sales brochure.

Second, consumer advertisements often appeal to the buyers' emotions to make them want the product. But the purchase of a pump, compressor, or ball bearing is rarely an emotional decision, and hard facts, not cute or clever presentations, are what the industrial buyer wants. LAP has found that 82 percent of trade magazine readers are looking for specific, detailed technical information in advertisements.

Third, consumer advertising sells, industrial or trade journal advertising *presells*. By this we mean that industrial advertising tells potential customers about your company and products before the sales representative calls.

A trade journal advertisement generates sales leads and creates awareness by presenting important information in an interesting, persuasive fashion. Like consumer advertisements, trade journal advertisements follow the basic format of headline, visual, and body copy.

Here are four things you should know before you sit down to write industrial advertising:

1. *Know your product.* The industrial buyer wants to know how it works, what it costs, what it can do, what it cannot do. Your advertisement should tell the reader, ''Here is how buying this product can benefit you and your company.''
2. *Know your competition.* There must be some way in which your product, service, or company is different from and better than the competition. Find your advantage, and tell your readers about it.
3. *Know your audience.* Are you writing for scientists? engineers? managers? purchasing agents? Know your readers by their job titles and functions, industries, and geographic locations.
4. *Know your purpose.* What is the *purpose* of producing and running the advertisement? Define your theme and message. Determine the kind of response you wish to generate. Create advertising that helps achieve specific marketing and sales objectives.

In addition to trade journal advertisements, industrial copywriters get involved with publicity, trade shows, newsletters, direct mail, and a variety of other promotions.

Now that you've learned what kinds of publications technical writers handle, the next five chapters will show you how to write and produce these documents.

3

How to Write Numbers, Units, Equations, and Symbols

Numbers, units of measure, mathematical equations, and symbols are used far more frequently in engineering and scientific writing than in ordinary composition. Because they communicate much of the data in any technical document, they must be written clearly and with consistency.

This sounds simple enough—until you remember that there is more than one way to write a number. For example, does that transmitter in your communications system weigh three-quarters of a ton, ¾ ton, 0.75 ton, .75 ton, $^{75}/_{100}$ ton, 1500 pounds, 1500 lb, or 1500 lbs? Did it cost twelve and a half million dollars, $12½ million, $12.5 million, 12.5×10^6 dollars, or 12.5 million dollars? Can you write these numbers any way you please? Or is there a well-defined format for writing mathematics?

As with grammar, spelling, and punctuation, there are set rules for writing numbers, units, equations, and symbols. Some are rigid; others vary from one style manual to the next. In this chapter, we combine the most universally accepted rules with plain common sense; the result is the following set of guidelines for writing mathematical terms in a correct, consistent style.

NUMBERS

Lord Kelvin, creator of the Kelvin scale of temperature, said, "When you can measure what you are speaking about, and express it in numbers, you know something about it." Unfortunately, few writers take the time to write numbers correctly, and more typos, errors, and inconsistencies occur in the mathematical portion of most technical documents than in the prose portion. By sticking to these rules, you can eliminate most of those mistakes:

Rule 1. Write out all numbers below 10.

The numbers one through nine are written out in English. You will find this rule the same no matter which grammar text or corporate style manual you consult. For example, you'd write

> nine tractors
> one trial run
> five command centers

The exceptions to this rule are numbers used with units of measure, time, dates, page numbers, percentages, money, and proportions, such as

> 2 yards
> 9-second delay
> 1 pound
> 6 years old
> 2 pm
> October 7, 1981
> page 8
> 4 percent
> $3 (not $3.00)
> 1:70
> 5 to 2 odds

Any number *greater* than nine usually is written in numerals:

> 10 times the amount
> 6,000 cells
> 209,000,000 people

There are many different ways to write a given number:

209,000,000
209 million
2.09×10^8
209×10^6
two hundred and nine million

Unfortunately, not all sources agree on how to handle these large numbers. We have adopted those rules which make for a clear, readable document.

Numbers in the thousands should be written with the separating comma; this makes them easier to understand at a glance.[1] (Note that some sources prefer to omit the comma in four-digit numbers.) The numbers in the left column should be rewritten in the form shown in the right column:

3000	3,000
149987	149,987
12000	12,000
123456	123,456
150000	150,000

Numbers in the millions can be written in two ways: as numerals or as a numeral plus the word *million*.[2]

3,500,000 or 3.5 million
150,000,000 or 150 million
15,000,000 or 15 million
1,500,000 or 1.5 million
56,887,423
1,234,567

[1] In Europe, a period is used in place of the comma. Thus, European technical writers would write 3,000 as 3.000, and 149,987 as 149.987.

[2] In writing amounts of money, however, we use the word *million* even if the amount of money exceeds 100 million. The figures in the left column should be rewritten as in the right column

$67,000,000	$67 million
$6,700,000,000	$6,700 million
six thousand million dollars	$6,000 million

As you can see, the numeral-and-word style in the right column is extremely clear and readable. It would seem only natural to use it for writing numbers in the billions, but we don't. The reason is that *billion* means a thousand million in the United States but a million million in most other countries. Since most technical documents reach an international audience nowadays, the term should be avoided altogether. After the million, then, it's back to numerals:

> 1,500,000,000
> 67,000,000,000

Numbers above the billions are also written in numerals, not as numerals and words, because the terms for these huge numbers—*trillion, quadrillion, quintillion,* etc.—are not well known to the average reader:

> 1,500,000,000,000
> 69,450,333,900,540,309

Another method for writing numbers, *scientific notation,* employs multiplication and exponents. In scientific notation, 1.5 million is written as 1.5×10^6, which means 1.5 times 1,000,000, or 1,500,000. Other examples are:

> $1.5 \times 10^9 = 1,500,000,000$
> $1.5 \times 10^8 = 150,000,000 = 150$ million
> $1.5 \times 10^7 = 15,000,000 = 15$ million
> $10^3 = 1,000$
> $7 \times 10^1 = 70$

Scientific notation is helpful when writing extremely large numbers as numerals become too clumsy:

> $1,000,000,000,000 = 10^{12}$
> $602,300,000,000,000,000,000,000 = 6.023 \times 10^{23}$

As a rule, it is best to use numerals whenever possible. Scientific notation may be compact, but its format is difficult to type and is likely to be reproduced incorrectly by graphic artists, typesetters, and printers. Besides, it's easier for the reader to think in terms of 100 billion, 7.5 million, and 12,000 than 10^{11}, 7.5×10^7, and 1.2×10^4.

Rule 2. Place a hyphen between the number and the unit of measure when they modify a noun.

A popular physics text poses this problem:

What upward force must be exerted on a 100 lb weight to cause it to fall with an acceleration of 10 ft/s²?

A physicist would quickly calculate the upward force to be 69 pounds. An English major, however, would know that the author of this textbook should have written *100-lb weight* instead of *100 lb weight*. Whenever a number and unit of measure are compounded to form an adjective, they are separated by a hyphen:

> 6-inch-diameter orifice
> 2-week-old culture
> 15,000-volt charge
> 8-pound baby

Note, too, that *all* numbers with units of measure are written as numerals, as we stated in rule 1. The hyphen separating the number and unit of measure is one of the most frequently omitted marks of punctuation in all technical writing. If you want your writing to be correct grammatically as well as technically, you must get in the habit of using it.

Rule 3. Decimals and fractions are written as numerals.

Decimals and fractions should be written as numerals; this form is more concise and readable than when they are written as words.

zero-point-three-four	0.34
five-point-five-eight-nine	5.589
three-fifths	⅗ or 0.6
four-tenths	⁴⁄₁₀ or 0.4

We recommend writing fractions as decimals whenever possible. Standard typewriters can type only ½ and ¼ properly; all other fractions must be constructed with the stroke or slash mark (/): 1/3, 1/5, 1/8, 3/16. Decimals are easier to type in a uniform fashion and are less likely to be reproduced incorrectly:

¼	0.25
34/100	0.34
6/20	0.3
½	0.5

A zero is always placed before the decimal point in numbers less than 1:

.34	0.34
.5000	0.5000

Writing a number as a decimal implies precision to the last decimal place—*0.50 cup of coffee* is coffee measured to the nearest hundredth of a cup. If a number is merely an approximation ("half a cup of coffee" is roughly half a cup, more or less), do not give it an inflated degree of accuracy by writing it as a decimal.

So we come to our next topic, approximations.

Rule 4. If a number is an approximation, write it out.

If you measure ³⁄₁₆ inch with an ordinary ruler, obviously your measurement is accurate to a sixteenth of an inch—and *not* to the ten-thousandth of an inch that writing the fraction ³⁄₁₆ as its decimal equivalent (0.1875) implies. Whenever you write a number as a decimal, you are stating that the number is accurate to the last decimal place. By writing an approximation as a decimal, you imply a degree of precision that is not there. Therefore, if a number is an approximation, write it out:

> half a glass of water
> a quarter of a mile down the road
> a third of the energy requirements

Distinguishing an approximation from a precise value is fairly easy for the experienced reader: Approximations usually are signaled by "hedge" words such as *about, up to, almost, more or less, roughly, approximately*, and *on the order of*. When you see these phrases, you know the numbers following them are not precise.

Rule 5. When many numbers are presented in the same section of writing, write them all out as numerals.

The rule here is to write all numbers as numerals when two or more numbers are presented in the same section of writing, as in this familiar Christmas ditty:

> On the 12th day of Christmas, my true love sent to me: 12 drummers drumming, 11 lords-a-leaping, 10 ladies dancing, 9 pipers piping, 8 maids-a-milking, 7 swans-a-swimming, 6 geese-a-laying, 5 gold rings; 4 calling birds, 3 French hens, 2 turtle doves, and a partridge in a pear tree.

Writing numbers in a consistent format makes them easier to read and compare. It looks neater, too. Here is an example from a technical proposal:

> The full-scale system contains 15 pumps, 5 fans, 5 ducts, and 3 heat exchangers.

There is one exception to this rule: If none of the numbers in the sections is greater than nine, write them all out as words:

> The pilot-plant system contains five pumps, one fan, one duct, and two heat exchangers.

Rule 6. Do not begin a sentence with numerals.

Readers, like drivers, become used to certain traffic signals. Periods, exclamation points, and question marks are red lights that signal the end of a sentence. A capitalized word is the green light that signals the start of a new sentence.

Although it is sometimes difficult to avoid, never begin a sentence with a numeral: it confuses the reader. Either you can write out the number, or, if the spelled-out number is unwieldy, rewrite the sentence so that it doesn't begin with a number.

2,000 test subjects participated in the experiment.	Two thousand test subjects participated in the experiment.
$250,000 went into the preparation of this proposal.	This proposal cost $250,-000 to prepare.

| One hundred fifty-four thousand six hundred and twelve soldiers will go through our training program in the next 10 years. | In the next 10 years, 154,612 soldiers will go through our training program. |

Rule 7. When one number appears immediately after another as part of the same phrase, one of the numbers (preferably the shortest) is spelled out.

If both numbers in the same phrase are written as numerals (left column below), the text seems to run together and is unclear at first glance. On the right, the same phrase is made easier to read by spelling out one of the numbers:

11 60-ohm resistors	eleven 60-ohm resistors
eight three-inch wrenches	eight 3-inch wrenches
three five-person teams	three 5-person teams
12 21-module software packages	twelve 21-module software packages
45 4,500-component radar systems	forty-five 4,500-component radar systems

UNITS OF MEASURE

Rule 8. Keep all units consistent.

Units of measure must be consistent. You can no more mix miles, meters, and micrometers than you can apples and oranges. By seesawing between various sets of units, the careless technical writer forces the reader to become a living conversion table.

Do not make your reader do extra work. Pick a set of units and stick with it throughout the text, tables, and illustrations in your document.

Length, mass, time, and other physical characteristics can be expressed in two basic systems of units: English and metric. Many U.S. scientists and engineers prefer the English system; it measures length in feet or inches, mass in slugs, force in pounds, time in seconds, and pressure in pounds per square inch.[3]

There are two different versions of the metric system. The older version is the centimeter-gram-second (cgs) system. As the name implies, it measures length in centimeters, mass in grams, and time in seconds. The more modern version is the Système International (SI) form of units. In SI, length is measured in meters, mass in kilograms, force in newtons, time in seconds, and pressure in pascals. These three systems—English, cgs, and SI—are compared below:

Quantity	English Unit	cgs Unit	SI Unit
time	second	second	second
mass	slug	gram	kilogram
force	pound	dyne	newton
length	foot	centimeter	meter
pressure	pound per square inch	dyne per square centimeter	pascal
energy	foot-pound	erg	Joule

Metric has become an international standard; only the United States, South Yemen, Burma, and Brunei still adhere to the English system. In this country, we have a voluntary metric conversion program, and many companies and institutions—including Sears, JC Penney, Chrysler, Ford, GM, IBM, Xerox, Levi Strauss

---

[3] The slug is a unit of mass not well known to the average person.

Mass and weight (a force) are often confused. Mass is the fundamental measure of matter; mass is that property that causes matter to resist acceleration.

Force, according to Newton's second law of motion, is mass multiplied by acceleration. Weight is the mass times the acceleration imparted by gravity.

The slug and kilogram are units of mass, the pound and Newton, units of force (and weight). Strictly speaking, pounds and kilograms are not interchangeable, since the former is force and the latter is mass. But, under standard conditions of Earth gravity, 0.4536 kilogram always weighs one pound, and both the pound and the kilogram are commonly (if not accurately) used as measures of weight. The slug, accepted in the scientific community, is practically unheard of in everyday usage.

& Company, Travelers Insurance, and NASA—are using the metric system in their product specifications and publications.

Of the two versions of the metric system, SI is fast living up to its name. Most major corporations and professional societies are adopting this system as a standard. These include the American Institute of Chemical Engineers, Technical Association of the Pulp and Paper Industry, Westinghouse Electric Corporation, Boeing Aerospace Company, the Institute of Electrical and Electronics Engineers, and the American Society for Testing and Materials (ASTM). ASTM publishes a widely accepted guide to SI units, *ASTM Standard for Metric Practice* (Document E 380-79, American Society for Testing and Materials, 1916 Race Street, Philadelphia, PA 19103).

Rule 9. Write units of measure as words or symbols; do not use abbreviations.

Appendix B lists SI units and their proper symbols. Never use any abbreviated form of a unit other than its accepted symbol. You can write time as "second" or "s" but not the informal abbreviation "sec." Likewise, current can be 0.6 ampere or 0.6 A, but not 0.6 amp.

3-sec delay	3-second delay *or* 3-s delay
12-hr hike	12-hour hike *or* 12-h hike

There is no set rule that says when to write units as words and when to write them as symbols. The choice of writing the unit as a word or symbol is up to you. Let common sense be your guide, and write units in whichever form is easiest to read. But be consistent, and use the same forms throughout the writing.

Experience has taught us to observe the following guidelines:

For single units, write out the unit.

> 12-inch ruler
> 6-second delay
> 24 hours
> 90-ohm resistor
> 18 meters

As we discuss in Chapter 4, it is better to use too few rather than too many abbreviations and symbols. Despite their official status, symbols for units such as seconds, hours, meters, and inches (s, h, m, in) look strange and unfamiliar to many readers. To avoid confusion, it's best to write out units whenever possible.

For units formed by multiplying and dividing other units, write units as symbols. When writing out multiple units becomes too cumbersome, simplify by using symbols.

32 feet per second squared	32 ft/s^2
667 cubic meters per second	667 m^3/s
500 joules per kilogram-degree kelvin	500 J/(kg·K)

Rule 10. Multiplication of units is indicated by a raised dot (·), a division by a slash.

To indicate the multiplication of units written as symbols, use a raised dot (·), not a multiplication sign (×) or hyphen (-). Indicate division with a slash, also known as a stroke or solidus (/).

0.3 J per mole × K	0.3 J/(mol·K)
32 ft per s^2	32 ft/s^2
kg per m-s^2	kg/(m·s^2)

If, however, you write out a unit in words, not symbols, indicate multiplication with a hyphen and division with the word *per:*

Top speed is 60 miles per hour.
The force was 240 kilogram-meters per second squared.
Typical plants process 80 million cubic feet per day of product.

Rule 11. If the number is one or a fraction of one, the unit is written in the singular.

Plural means more than one. Yet a respected trade journal recently reported an experimental plant as processing *0.8 tons of raw material per hour*. Something seems amiss here: we know that 0.8 is equal to ⅘. Is there such a quantity as four-fifths *tons?* Surely the writer should have written *0.8 ton,* and not *0.8 tons*.

Describing a fraction of one as being plural (0.8 tons, 0.5 operators, 0.25 seconds) is an easy mistake to make—these constructions often sound right. But they are incorrect: no number of one or less can be described as plural. You can have 0.25 or a quarter of a second, but certainly not a quarter of a seconds.

>1.56 tons
>0.56 ton
>0.50 ton
>½ ton
>half a ton
>20 tons

The number 0 is an exception to this rule. Although 0 is certainly less than one, the plural form of the unit is used.

>zero dollars
>0 pounds

There are, however, no plural versions of the *symbols* used for units of measure. (Pounds, for example, cannot properly be written as *lbs.*) Therefore, symbols of units are *always* written in the singular.

>45 lb
>12 in

Rule 12. Write secondary units in parentheses after the primary units.

Recently, an engineer decided to write a paper for a professional journal. The journal sent her a booklet of author's guidelines which instructed her to use SI units.

"Ridiculous!" complained the engineer. "American engineers—at least all the ones *I* know—think in terms of inches, pounds, feet, and psi, not kilograms and meters. If I say my boiler steam rate is 65,000 lb/h, people will know what I'm talking about.

But if I write it in metric, well . . . no U.S. engineer has a feel for how big an 8.2-kg/s flow rate is. . . .

"Why do they insist on this SI baloney? Everyone who reads my paper will just convert to English units anyway!"

There's some truth to her complaint. We do think in terms of the units of measure we grew up with. (If you don't believe it, ask the next 10 people you meet how much they weigh. How many answer in kilograms instead of pounds?)

To satisfy your publisher and keep your readers happy at the same time, you can write a document in *two* sets of units. The secondary unit is written in parentheses and follows the primary unit. If you use two sets of units, handle them consistently throughout the writing. The example on the left mixes English and metric units indiscriminately. At right, it's rewritten in a consistent format: primary units in metric followed by secondary units in English.

Use a 10-meter section of 2-inch (5.08-centimeter) diameter pipe.

Use a 10-meter (32.8-foot) section of 5.08-centimeter (2-inch) diameter pipe.

EQUATIONS

Mathematical equations frequently communicate more effectively and more eloquently than words; the most well-known scientific thought in Western civilization is probably $E = mc^2$.

To typists, editors, graphic artists, and printers, however, these cumbersome, complicated formulas are a production nightmare. To begin, standard typewriters cannot make the superscripts, subscripts, brackets, arrows, mathematical signs, Greek letters, and other symbols used in equations. Worse yet, these strange-looking marks can be "Greek" to the people involved in editing and producing the manuscript. Their lack of familiarity with the author's specialized notation often results in equations that are incorrectly typed and reproduced in the final document. As you know, even something as innocent as a misplaced subscript can completely change the meaning of a mathematical expression.

Fortunately, these mistakes can be virtually eliminated with a little extra care and patience. First, do not type equations on your first draft—write them in longhand as clearly as possible. Trying to improvise on a standard typewriter will only frustrate and confuse both author and editor. In its final form, the equation will be typed, typeset, or penned in neatly by hand. (Chapter 7 covers the mechanics of producing technical documents in great detail.) Either way, authors *must* proofread their own work to make sure the equations are reproduced correctly before their manuscripts are printed and distributed.

Equations will never be fun to write. But by following these hints and the guidelines below, you can handle them properly and with a minimum of personal agony.

Rule 13. Center and number equations on a separate line in the text.

Equations are often lengthy and unwieldy and take up more than one line of the text. Most are simply too big to handle within the text itself.

The general first-order linear equation is $dy/dx = p(x)y + q(x)$ and the general second-order linear equation is $d^2y/dx^2 = p(x) \, dy/dx + q(x)y + r(x)$.

As you can see, it's difficult to run these equations in a single-space text. Everything is much clearer and cleaner if we write each equation on a separate line:

The general first-order linear equation is

$$\frac{dy}{dx} = p(x)y + q(x) \tag{1}$$

and the general second-order linear equation is

$$\frac{d^2y}{dx^2} = p(x)\frac{dy}{dx} + q(x)y + r(x) \tag{2}$$

Note that the equations are numbered in the order in which they appear. This makes it easy to refer to them later in the text.

We can rewrite the general first-order linear equation (Eq. 1) as

$$\frac{dy}{dx} + p(x)y = q(x) \tag{3}$$

Centering and numbering equations make the document neater and more readable. Short equations, however, may be run into the text if you like.

On Ludwig Boltzmann's gravestone is carved his formula for entropy, $S = K \log W$. Boltzmann committed suicide in 1906 at the age of 62.

Rule 14. Keep all division lines (or fractions) and plus or minus signs on the same level.

Once we had a secretary who approached technical typing assignments with the same enthusiasm that most vegetarians have for a good 12-ounce steak. Here's a sample of her work:

$$\Delta P_t = (4/\pi)(Ne\text{Re}_D)\frac{n}{D^4}\frac{V(L)}{D}$$

To make sure your equations do not wander across the printed page like a sine wave, keep all plus, minus, multiplication, and division signs aligned with the equals sign.

$$\Delta P_t = \frac{4}{\pi} Ne\text{Re}_D \frac{n\dot{V}}{D^4}\left(\frac{L}{D}\right)$$

$$A = B + \frac{C}{D} + \frac{E}{F}$$

$$x = e^{-bt}(C_1 \cos\theta + C_2 \sin\phi) + \frac{F_0}{K_1}$$

$$\frac{dy}{dx} = p(x)y + q(x)$$

As an exercise, try typing that last equation so that the division line is perfectly aligned with the equal sign. You will find it hard to do; as we mentioned, most standard typewriters are inadequate for this kind of work.

For a series of equations, write them so that their equals signs are aligned vertically:

$$x = A + B + C$$
$$y = D + E$$
$$z = F + G$$

Rule 15. Punctuate words used to introduce equations just as you would words forming part of any sentence.

Like any other phrase, an equation is part of your sentence. Why, then, do writers constantly introduce formulas with punctuation that does not fit the sentence structure? A high school physics student wrote in a laboratory report:

The current in the wire is calculated by using Ohm's law.

This sentence is properly punctuated. It would be grammatically incorrect to add a colon.

The current in the wire is calculated by using: Ohm's law.

Clearly, the colon is out of place. The construction should not change if Ohm's law is written as an equation. Yet many authors would write the following:

The current in the wire is calculated by using:

$$E = IR$$

There is no logical reason for using punctuation incorrectly whenever equations are presented. Punctuate the words introducing equations as you would any ordinary phrase, and use no punctuation following displayed equations:

The current in the wire is calculated by using

$$E = IR$$

Many writers and editors unfamiliar with the language of mathematics seem to be intimidated by equations. If you fit this category, it might help you to mentally substitute the phrase *this equation* for the actual formula when you punctuate sentences. The sentence structure will be the same, but thinking in terms of words instead of numbers and symbols should make things easier. Here are a few more examples:

A straight line is given by:
$$y = mx + b$$
On the other hand
$$P = \frac{F}{A}$$
The distance formula is,
$$d = rt$$

A straight line is given by
$$y = mx + b$$
On the other hand,
$$P = \frac{F}{A}$$
The distance formula is
$$d = rt$$

SYMBOLS

Rule 16. Use too few rather than too many symbols.

Symbols provide a way of writing units of measure, mathematical variables, physical constants, and other terms in an abbreviated fashion. They can reduce writing and typing time and make sentences briefer. A few examples of words and their symbols are presented in the table below:

Word	Symbol
percent, percentage	%
degree	°
addition	+
dollar	$
mercury (the element)	Hg
Mercury (the planet)	☿
snow	✳
gas	↑
ohm	Ω
greater than	>
square root	√
cent	¢
flat note (music)	♭
Pisces (zodiac)	♓
Silurian soil	S

Some of these symbols (%, +, $, ¢) are probably familiar to you; you may not recognize others (♀, ✳, ↑). We can assure you that you will never write for a reader who knows the meanings of *all* the hundreds of symbols used in technical literature. Therefore, if there is the slightest chance that any part of your audience will not recognize a symbol, write out the word instead. Avoid confusion and misunderstanding by using too few rather than too many symbols.

Even readers in your specialized field may not know all the symbols you do. If you are a chemist writing for other chemists, it's a safe bet that writing H_2O instead of *water* will not confuse anybody. But

might give some readers a headache. The writer would have been better off simply writing out the name of the compound—in this case, acetylsalicylic acid (aspirin).

However when they are properly used, symbols, like any abbreviation, can make your sentences shorter and more readable. Especially in equations, you can see how substituting symbols for words make the writing concise.

force equals mass times
acceleration $F = ma$

energy equals mass times
the speed of light squared $E = mc^2$

Rule 17. Define the symbols you use.

Make sure your reader knows what you are talking about by defining the symbols you use. Clearly identify each symbol as it is introduced in the text. Then readers do not have to refer to an appendix, a glossary, or a technical dictionary to understand your meaning.

The most acidic compound known is $HClO_4$ (perchloric acid).

To convert Celsius to Fahrenheit, use

$$°F = 1.8(°C) + 32°$$

where F is degrees Fahrenheit and C is degrees Celsius.

In addition to defining symbols in the text, a list of nomenclature placed at the end of the document is always a handy reference. The nomenclature section lists, in alphabetical order, the symbols used, along with their definitions and units of measure.

Nomenclature

Symbol	Definition	Unit
B	magnetic induction	weber per meter squared
C	capacitance	farad
E	electric intensity	volt per meter
L	inductance	henry
R	resistance	ohm

This is very helpful for readers going through the actual calculations in your paper or report.

Rule 18. Avoid duplication of symbols.

Electric resistance, degrees Rankine, the universal gas constant, and the mean radius all happen to be symbolized by R. In the same way, T stands for the prefix *tera* (10^{12}), for temperature, and for the tesla, a unit of magnetic flux.

A given symbol frequently stands for more than one variable or unit of measure. This is because there are not enough symbols to go around, and technical experts in one discipline will adopt a symbol regardless of whether it's used in another area of science.

You may find yourself using several terms, all represented by a single symbol. Obviously, writing these terms as symbols would cause confusion, and duplication of symbols must be avoided. Either write out all terms as words or redefine them with new, non-standard symbols.

The R in the test wire was 60 ohms; ambient temperature was 520°R.

The resistance in the test wire was 60 ohms; the ambient temperature was 520°R.

The r in the test wire was 60 ohms; the ambient temperature was 520°R. (Here r was defined to be resistance at some point in the text.)

Rule 19. Since symbols are substitutes for words, they must fit grammatically into the structure of the sentence.

The philosopher, J. J. C. Smart, once wrote the following:

If the argument is valid, that is, if r really does follow from p and q, the argument will lead to agreement about r provided that there is already agreement about p and q.[4]

Philosophy and logic aside, you will note that Smart has fit his symbols (r = conclusion, p = first premise, q = second premise) neatly into the structure of his writing. The sentence, although wordy, is properly punctuated.

Symbols are not meaningless, alien markings; they are substitutes for words. And so, like the words they represent, symbols must be handled with the proper rules of grammar and punctuation.

[4] In George I. Mavrodes (ed.), *The Rationality of Belief in God*, Prentice-Hall, Englewood Cliffs, N.J., 1970, p. 97.

A Few Useful Rules of Grammar, Abbreviation, and Capitalization

There are hundreds of rules of grammar, so we selected a few which we feel illustrate points that come up again and again in technical writing. The rules include the correct use of hyphens, the serial comma, parentheses, and capitals as well as the avoidance of dangling participles, run-on sentences, and sentence fragments.

GRAMMAR

Rule 20. Two words compounded to form an adjective modifier are hyphenated.

This is an iron-clad rule! The two words *iron* and *clad* form an adjective modifying *rule*. However, a compound such as *up to date* as in the sentence "Bring him up to date" is not an adjective modifier and so is not hyphenated. Whenever two or more words are compounded to form an adjective, they are hyphenated.

space time continuum	space-time continuum
state of the art technology	state-of-the-art technology
long range, high power radar	long-range, high-power radar

Hyphens are used in compound modifiers because they help the reader avoid confusion. For example, if a compound such as *first-*

49

order reaction were *not* hyphenated, the reader would expect *first* to modify the phrase *order reaction*; when the hyphen is included, the reader sees a *reaction of the first order*.

No hyphen is needed in adverb-participle combinations if the adverb ends in *ly*.

technically accurate manual
artificially induced sleep
financially stable organization

Note: Adjectives ending in *ly* are hyphenated when they are used with participles, as in *friendly-sounding voice*.

Many compound words that are hyphenated *before* a noun are not hyphenated when they occur *after* the noun.

This is my up-to-date report.	*Up-to-date* precedes the noun and is hyphenated.
Bring them up to date.	*Up to date* follows the noun and is not hyphenated because it does not modify a noun.

To avoid confusion, a modifier that would be hyphenated before a noun retains the hyphen when it follows a form of the verb *to be*.

The physicist is theory-oriented.
The research report was second-rate.

Do not hyphenate scientific terms, chemicals, diseases, and plant and animal names used as unit modifiers if no hyphen appears in their original form. *This is the major exception to rule 20.*

sulfur dioxide emissions
swine flu epidemic
an apple tree grove

Rule 21. Two adjacent nouns are hyphenated if they express a single idea and if, without the hyphen, this idea is not immediately clear.

In English, the trend has been first to join two nouns with a hyphen and then, after they have become accepted, to make them a single word.

air craft	air-craft	aircraft
type setting	type-setting	typesetting

The hyphen is used to make the connection between the two words immediately obvious to the reader. For example, the un-hyphenated combination *feed pipe* seems, at first reading, to be an imperative to feed a hungry pipe. With the hyphen, we instantly recognize the compound *feed-pipe* as the pipe in which fluids feed into a process system. Some other examples of how a hyphen clarifies the idea in a compound are shown below:

safety valve	safety-valve
light year	light-year

The dictionary is the final authority on whether two words are separate, hyphenated, or joined into one word.

Rule 22. In a series of three or more terms with a single conjunction, use a comma after each except the last.

The comma after the next-to-last term (the term that comes just before the conjunction) is known as the *serial comma*.

Today, writing authorities are divided as to whether the serial comma should be used. The major reference works on English grammar recommend using this comma. We agree; it can add clarity and make sentences flow more smoothly. Therefore, you should write:

The four most abundant elements in the earth's crust are oxygen, silicon, aluminum, and iron.

Reports, proposals, and manuals are the responsibility of the technical writing department. The advertising department handles brochures, catalogs, and press kits.

She is a communications consultant for the three major television networks: ABC, NBC, and CBS.

Rule 23. If a sentence contains an expression within parentheses, the expression is punctuated as if it stood by itself, except that the final stop is omitted (unless it is a question mark or exclamation point).

In technical writing, sentences often end with an expression within parentheses. Omit the final period in this parenthetical expression. Punctuate the sentence outside the parentheses exactly as if the parenthetical expression were not there.

Incorrect	*Correct*
The growth rate has increased by 12 percent each year (see Fig. 8.).	The growth rate has increased by 12 percent each year (see Fig. 8).
The growth rate has increased by 12 percent each year (see Fig. 8.)	

The rule also applies to parenthetical expressions that appear in the middle of a sentence.

It is extremely unlikely that two people (barring identical twins) share precisely the same gene combinations.

At the science fair, the chemist (or was he more of an alchemist?) turned red wine into milk before our eyes.

Rule 24. Avoid dangling participles.

A verb ending in *ing* is called a *present participle*. Grammarians say a participle is *dangling* when it is attached to the wrong subject in a sentence. Consider this example:

Turning over our papers, the chemistry examination began.

This sentence says that the chemistry examination began after it turned over our papers! Here, the participle *turning* is dangling because it is attached to the wrong subject: the chemistry examination. It should refer to the *we* that did the actual turning over of the papers. The correct form is

Turning over our papers, we began taking the chemistry exam.

Dangling participles occur frequently in technical writing, sometimes resulting in absurd sentences. Sentences with dangling participles must be rewritten to make their meaning clear.

King Tut's tomb was unearthed while digging for artifacts.	Archaeologists unearthed King Tut's tomb while they were digging for artifacts.
Developed by scientists at the Atlanta Disease Control Center, Dr. Jones asserted that the serum was a wonder drug.	Developed by scientists at the Atlanta Disease Control Center, the serum was hailed as a wonder drug by Dr. Jones.
Sitting serenely in the laboratory, the apple tasted great to her.	Sitting serenely in the laboratory, she enjoyed the great taste of the apple.

Rule 25. Avoid run-on sentences.

A *run-on sentence* is a sentence that runs on and on without a proper pause. It loses the readers in an unbroken string of independent clauses and leaves them breathless and confused.

Strictly speaking, a run-on sentence is defined as two or more *independent clauses* (parts of sentences that can stand alone as separate sentences) with either a comma or no punctuation at all between them.

The vacuum tube burned out it will be replaced.
The computer is down, you must fix it soon.

These sentences are punctuated incorrectly. When two or more independent clauses not joined by a conjunction (*and, or, but, for*) are to form a single sentence, they must be separated by a semicolon.

The vacuum tube burned out; it will be replaced.
The computer is down; you must fix it soon.

Run-on sentences also can be broken into two separate sentences.

The vacuum tube burned out. It will be replaced.
The computer is down. You must fix it soon.

Or, you can join the independent clauses with a comma followed by a conjunction.

The vacuum tube burned out, but it will be replaced.
The computer is down, and you must fix it soon.

Run-on sentences can muddle your meaning, and their use shows poor knowledge of grammar. Adding the proper punctuation or forming several shorter sentences makes your writing clear and correct.

Rule 26. Avoid sentence fragments.

A complete declarative sentence must contain a subject and a finite verb. A sentence missing an essential element (subject, verb, or object) is an incomplete sentence, or *sentence fragment*.

The omission of a verb or subject is *usually* an error born of laziness. The result is a nonsentence that makes no sense.

Maxwell's remarkable discovery that the speed of propagation of electromagnetic effects is precisely the same as the speed of light in the same medium.

Well, what *about* Maxwell's discovery? The lack of a verb makes this fragment incomplete, grammatically incorrect, and extremely frustrating. No statement was made, no idea communicated. But add a simple verb and the words suddenly make sense.

Maxwell *made* the remarkable discovery that the speed of propagation of electromagnetic effects is precisely the same as the speed of light in the same medium.

Sometimes, skilled writers use short sentence fragments to achieve added impact and smooth the flow of the writing.

At Venture Electronics, we are fast. And reliable.

The new fluidized bed combustion boiler should increase heat transfer efficiency by up to 500 percent. *If* the system works.

Sentence fragments should be used sparingly. Too many departures from standard written English make readers uneasy and give the impression that the writer is uncertain of the rules of grammar.

ABBREVIATION

Rule 27. Avoid abbreviations in writing, and use too few rather than too many.

Abbreviations proliferate in our technical society. People are as fond of the shorthand of abbreviations as they are of jargon, and it's no wonder that two of the characters in the futuristic film *Star Wars* are known simply as R2D2 and C3PO.

Everything would be fine if we could guarantee that our readers would understand all the abbreviations we use, *but do not assume that they do*.

You may know what OPEC, NASA, GE, and USSR stand for, but not everyone does. A college student may know what MIT, RPI, and UVM stand for, but these initials do not communicate meaning to everyone. Even the often-used TGIF (thank God it's Friday) isn't universal.

When we rely on abbreviations, we start a process of "inbreeding" that may prevent us from clearly communicating our thoughts to people outside our discipline or department. Government employees may understand that DOT means Department of Transportation or that OMH means Office of Mental Health, but they had better not rely on these abbreviations if they need to explain these departments to an outsider. In the same way, bureaucrats may be able to roll off abbreviations such as MRV, ICBM, LEM, and OAS, but they should not assume that laypeople will understand the meaning unless the words are spelled out first.

Explain every abbreviation you use. The first time you intend to abbreviate, write out the word, followed by the abbreviation in parentheses. In this way, you can be sure that all your readers are familiar with the abbreviation.

The integrated computer-aided system (CAS) is not a novelty.

One major manufacturer of electronic systems has used a CAS approach on all its major projects since 1968.

The only abbreviations that seem to catch on quickly, and can be used regularly with no explanation, are those such as DNA or LSD, where the abbreviations stand for extremely lengthy names—names that would not convey their meanings if they were spelled out.

Rule 28. Omit internal and terminal punctuation in abbreviations.

This is not a hard and fast rule, but mainly a matter of style and taste. Some organizations tell their editors to punctuate abbreviations; others prefer to delete the little dots. We choose to omit internal and terminal punctuation because this style has a cleaner, more readable appearance.[1] Thus, we abbreviate *inch* not as *in.* but as *in*, and we write *5 pm* instead of *5 p.m.* Likewise, write *US* instead of *U.S.* Addition examples are

R.S.V.P.	RSVP
p.s.i.	psi
U.F.O.	UFO

There is one major exception to this rule: The final period *is* used when the abbreviation spells out a whole word. But if this word is a unit of measure, the final period is omitted.

no. (number)
Fig. (figure)
lb (pound)
yd (yard)
mi (mile)

Rule 29. The abbreviation for a specific word or phrase takes the same case (upper case or lower case) as the word or phrase. An abbreviation that is an acronym usually is written in capital letters, except when the acronym stands for a unit of measure.

An abbreviation for a single word begins in the same case as the word itself.

Prof. Smith (Professor Smith)
etc. (et cetera)
cw (clockwise)

---

[1] The internal punctuation should be retained in proper names, however. Write J. J. Thompson, not J J Thompson.

ed. (edition)
Westinghouse Corp. (Westinghouse Corporation)

When an abbreviation is formed from the first or first few letters of each major word of a compound term, it is called an *acronym*. (Thus *ATC* is an acronym for *air traffic control*.) Usually acronyms are written in uppercase letters, regardless of whether the words they stand for begin with capitals.

VHF (very high frequency)
USA (United States of America)
OEM (original equipment manufacturer)
AM (amplitude modulation)

However, acronyms that stand for units of measure are always written in lowercase letters.

ppm (parts per million)
rpm (revolutions per minute)
mph (miles per hour)
bps (bits per second)

Rule 30. Avoid using signs in writing (″ for inch, ′ for foot).

Typewriters make life easy for the technical writer—sometimes *too* easy. It's tempting to press the key for double quote marks instead of writing out the word *inches*. Likewise, the apostrophe is a convenient stand-in for the word *foot*.

But this convenience invites confusion. A minus sign could be mistaken for a misplaced hyphen, dash, or underscore. The double quote marks immediately program readers for a quotation, not a measurement in feet. Even the symbol for *at* (@) can lead to confusion. So, simply avoid all unnecessary signs. When you write out the word, you opt for total clarity.

"A 15″ opening in the wall allows proper ventilation," said Dr. Jones.	"A 15-inch opening in the wall allows proper ventilation," said Dr. Jones.
Standard containers hold 75# of #5 carbon black.	Standard containers hold 75 pounds of no. 5 carbon black.

CAPITALIZATION

Rule 31. Capitalize trade names.

Every so often, a trademarked product comes into general use, and writers lose touch with the word's commercial origins. So, when someone inadvertently asks for a Kleenex instead of a tissue, the writer should understand that one of these words is a trade-marked name—not a generic name—and so must begin with a capital. Some product names have become English words in their own right, and they do not require a capital. For example, *fiberglass* is a word derived from Fiberglas.

Technical writers, as we have so often said, must uphold a high standard of accuracy. Don't write *I need to xerox this report,* because Xerox is the name of a company and of one *copier;* it is not a synonym for the word *photocopy.* To help you keep in mind some of the current trade names often confused with generic products or processes, here is a list of some that are, in a way, dangerously familiar.

Plexiglas	Xerox	Jell-O
Vaseline	Novocain	Frisbee
Scotch Tape	Valium	Frigidaire
Formica	Kodak	Univac
Polaroid	Band-Aid	Astro-Turf
Mace	Ping-Pong	Sanforized

Naturally, companies which have taken out a trademark on a product wish to protect the use of its name. Sanka is advertised as Sanka Brand coffee because its manufacturer does not want the public to get in the habit of allowing *Sanka* to become synonymous with all decaffeinated coffees. At one time, *aspirin* and *borax* were trade names, but today they have lost all association with their original manufacturers.

Rule 32. Do not capitalize words to emphasize their importance.

Someone once said, "Those who praise everyone do a vast dis-service to those few who truly deserve it." In the same way, the arbitrary capitalization of words to make them important or stand out can lead to confusion as well as cheapen those words that should be capitalized.

Advertising and publicity can enhance the Value Package of your product.	Advertising and publicity can enhance the value package of your product.
Certain kinds of fuels can cause Fuel Starvation as cells age.	Certain kinds of fuels can cause fuel starvation as cells age.
Burning is a Chemical Reaction in which oxygen atoms combine with the atoms of the substance being burned.	Burning is a chemical reaction in which oxygen atoms combine with the atoms of the substance being burned.

There are possible exceptions to this rule. A graphic designer may choose to use a nonstandard style of capitalization for headlines, subheads, titles, and captions. Although it is technically uncalled for, this gambit may lend a brochure or an advertisement a more pleasing, eye-catching appearance. Strictly speaking, for maximum clarity, do not abuse the rules of capitalization for an effect that might cause your reader any puzzlement.

Rule 33. Capitalize the full names of government agencies, departments, divisions, organizations, and companies.

Official names and titles are capitalized:

> Air Pollution Control Division
> U.S. Small Business Administration
> Spartan Engineering Company

Do *not* capitalize words such as *government, federal, agency, department, division, administration, group, company, research and development, engineering, manufacturing,* and *quality control* when they stand alone. They are capitalized only when they are part of an official name.

This is a problem for Research and Development, not Engineering.	This is a problem for research and development, not engineering.

| | This is a problem for the Research and Development Department, not the Engineering Department. |
| She is the head of her Division in the Company. | She is the head of her division in the company. |

Rule 34. Capitalize all proper nouns and adjectives unless usage has made them so familiar that they are no longer associated with the original name.

Science offers many rewards. One is the pleasure of having your discovery named after you. Hundreds of theories, laws, formulas, numbers, and units of measure have been named for the scientists who conceived of them. These words retain the capital letters used in the proper names from which they derive.

Kelvin scale (William Thompson Lord Kelvin)
Heisenberg uncertainty principle (Werner Heisenberg)
Mach number (Ernst Mach)

Repeated usage has made the more common proper nouns and adjectives so familiar that most people no longer associate them with the names of their founders. When this happens, the capital letter is dropped.

diesel (Rudolf Diesel)
ampere (Andre-Marie Ampere)
hertz (Heinrich Rudolf Hertz)
ohm (George Simon Ohm)

5

Principles of Technical Composition

Making your writing readable requires more than just an awareness of spelling, grammar, and syntax. It means developing a coherent, precise style. There are several valuable principles which guide technical writers in forging a lively, concise, and individual way of expressing ideas. Here are a few general rules which apply to technical (as well as nontechnical) writing.

Rule 35. Use the active voice.

In sentences written in the active voice, action is expressed directly; the subject is doing the acting. *John ate the pizza* is active because the subject, *John,* is eating the pizza. By comparison, *The pizza was eaten by John* is a passive sentence in that the action is indirect.

Whenever possible, use the active voice. Your writing style will be more direct and vigorous; your sentences, more concise. The passive voice sounds stiff and weak, as you can see in these examples:

Passive	*Active*
Dolphins were taught by researchers in Hawaii to learn new behavior.	Researchers in Hawaii taught dolphins to learn new behavior.
Control of the furnace is provided by a thermostat.	A thermostat controls the furnace.
Fuel cost savings were realized through the installation of thermal insulation.	The installation of thermal insulation cut fuel costs.
The instruction manuals are frequently updated by our technical editors.	Our technical editors frequently update the instruction manuals.

The passive voice *is* used, however, when the doer of the action is unknown or unimportant (or less important than the action itself).

The first smallpox vaccination was given in 1796.

It was found that a virus induces cells to make interferon.

Rule 36. Use simple rather than elegant or complex language.

In the foreword to his book *The Solid Gold Copy Editor,* Carl Riblet, Jr., has this to say about simplicity in writing:

> The copy editor believes that an interesting story should be remarkable for its simplicity. For example—he knows that rain falls in drops, and he is dedicated to the idea that to describe rain otherwise, as "precipitation, water droplets condensed from atmospheric water vapor and sufficiently massive to fall to the earth's surface," is unnecessary, undesirable and unendurable.[1]

Like the newspaper copy editor, the technical writer strives for simplicity by avoiding fancy phrases, bombast, and "purple prose." Simple language communicates more effectively than complex language, and communication—not literary style—is the mark of good technical writing.

[1] Carl Riblet, Jr., *The Solid Gold Copy Editor,* Aldine, Chicago, 1974, p. vii.

Complex	*Simple*
I feel that the application of these key principles will provide me with a clear-cut method of handling problem situations, while affording the employee the opportunity, the experience, and the feeling of interacting with me.	The application of these key principles will give me a clear-cut method of solving problems and allow me to work closely with the employee.
Another very important consequence of Einstein's theory of special relativity that does not follow from classical mechanics is the prediction that even when a body having mass is at rest, and hence has no kinetic energy, there still remains a fixed and constant quantity of energy within this body.	According to the theory of special relativity, even a body at rest contains energy.
The corporation deemed it necessary to terminate Joseph Smith.	Joseph Smith was fired.
The author hereby and irrevocably appoints Scott Lewis, Inc., as his sole and exclusive literary agent with respect to this and all future works.	Scott Lewis, Inc., is my agent.

Rule 37. Delete words, sentences, and phrases that do not add to your meaning.

Unnecessary words waste space and the reader's time, and they make strong writing weak. The fewer words you use, the better. After you have written a first draft, go through it with a pencil and strike out all words, sentences, phrases, and pages which do not add to your meaning. Here are a few examples of how concise writing is livelier and more readable than wordy prose:

Wordy	*Concise*
It is most useful to keep in mind that the term *diabetes mellitus* refers to a whole spectrum of disorders.	*Diabetes mellitus* refers to a whole spectrum of disorders.
Anthropologists have long observed that the Jalé people, who live and dwell in New Guinea, will exhibit cannibalism in that they eat the bodies of enemies they slay in the conflict of war.	The Jalé people of New Guinea eat the bodies of foes slain in war.
In the majority of cases, the data provided by direct examination of fresh material under the lens of the microscope are insufficient for the proper identification of bacteria.	Often, bacteria cannot be identified under the microscope.

As you can see, concise writing is crisper, simpler, and easier to read.

Sometimes deleting whole paragraphs can improve a piece of writing. Here is the lead paragraph from an article recently published in a leading trade journal:

It is both exciting and rewarding to discover that the scientific principles of one's profession can have immediate and gratifying expression in daily life. A case in point occurred recently, and I think it is appropriate to relate.

Beginning technical writers feel the need to ease the reader into their writing with this kind of lengthy "warm-up" introduction. But this paragraph does not contain news, or important facts, or items of interest. (Of *course* the author thinks "it is appropriate to relate"; otherwise, he or she would not have written the article.)

How much better to delete the unnecessary paragraph and plunge right into the story!

Rule 38. Use specific and concrete terms rather than vague generalities.

When engineers read a brochure or report, they seek detailed technical information—facts, figures, data, recommendations, and conclusions. Omitting technical detail for brevity's sake is an error that severely undercuts the value of your writing. By its very nature, technical writing must deal in specifics, not generalities.

Below are two versions of a technical advertisement. The one on the left was written by an advertising agency copywriter; it is clear and concise, but a little short on detail. A technical writer's rewrite appears at the right. This advertisement, though slightly longer, is much more persuasive because it deals in hard numbers—dollar savings, on-line availability, and efficiency.

How Our New Scrubber Means Big Savings for Samson Power

Aircom's new scrubbing system saves the Samson Power Company a fortune in fuel every day.

How?

By allowing Samson to burn cheap, high-sulfur coal instead of expensive compliance oil, the Aircom Scrubber cuts fuel costs.

This new system has proved itself efficient. And reliable.

To find out more, send for our free brochure.

Aircom's Scrubbing System Saves Samson $7,000 Every Day

With Aircom's new scrubbing system, Samson can burn high-sulfur coal instead of expensive compliance oil and *still* meet all federal and local emission control regulations.

The result is a fuel cost savings of $7,000 a day—a 35 percent reduction in Samson's annual fuel bill.

For over one year, the Aircom system at Samson has demonstrated an on-line availability of 98 percent. And an average SO_2 removal efficiency of 92 percent.

To find out how this commercially proven scrubber can clean the air and cut your fuel costs, send for our free brochure today.

Do not be content to say something is good, bad, fast, or slow when you can say *how* good, *how* bad, *how* fast, or *how* slow. Be quantitative, not qualitative, whenever possible.

He ran fast.	He ran the 100-yard dash in 10.2 seconds.
The sun is hot.	The sun is hot—almost 11,000°F at its surface.

The *words* you choose should be specific and concrete.

Our measurements are not precise because the experimental apparatus was in poor condition.	Our weight measurements are not precise because the scale was working poorly.
The expedition was delayed for a time because of unfavorable weather conditions.	The expedition was delayed one week because of snowstorms.

Rule 39. Use terms your reader can picture.

In *Heavy Equipment,* written, designed, and illustrated by Jan Adkins, the author describes his subject in these words:

> They are the big machines, the heavy equipment that chews at the earth and builds on it. They are so strong! The power of a thousand horses lives in their metal hearts and nothing can stand against them.[2]

Colorful language, when used properly and sparingly, can create in the mind a picture far more indelible than any catalog of facts and statistics. Technical writers should welcome the picturesque phrase when it helps get the message across. Here's how Russell F. Doolittle, author of the article "Fibrinogen and Fibrin," helps etch the concept of platelets into the reader's imagination:

> Prick us and we bleed, but the bleeding stops; the blood clots. The sticky cell fragments called platelets clump at the site of the puncture, partially sealing the leak.[3]

[2] The excerpt from *Heavy Equipment* was quoted in a book review in *Scientific American,* December 1981, p. 38.

[3] Russell F. Doolittle, "Fibrinogen and Fibrin," *Scientific American,* December 1981, p. 126.

The author could have written, "Platelets help our blood to clot," but by using vivid, concrete terms, such as *sticky cell fragments, clump, site of the puncture,* and *sealing the leak,* Doolittle has created a crisp, clear image of blood clotting, an image immeasurably preferable to a dry recitation of scientific technicalities.

Rule 40. Use the past tense to describe your experimental work and results.

Research reports are written in the past tense because they describe work which has been done. No one expects you to put laboratory experiments that occurred in the past into the livelier present tense; it would be inappropriate. So, writing which refers to an experiment will have the flavor of the past, as in the following sentences:

> Sensitivity after drug withdrawal began an average of a few days after the last dose and lasted an average of six days.

> The flow rate was measured for each of the three pipe lines.

> A mixing valve was used to create a liquid-liquid dispersion.

Rule 41. In most other writings, use the present tense.

Hypotheses, principles, theories, facts, and other general truths are expressed in the present tense. Avoid using the conditional *could* or *would,* and do not invoke the future tense needlessly.

The changing about of one amino acid in the chain could make an entirely different protein.	The changing about of one amino acid in the chain makes an entirely different protein.
Crystals would form from fusion if the temperature or pressure were high.	Crystals form from fusion at high temperature or high pressure.
The rocket will usually burn eight times its weight in liquid oxygen.	The rocket usually burns eight times its weight in liquid oxygen.

Rule 42. Avoid needlessly technical language. Make the technical depth of your writing compatible with the background of your reader.

If a technical term makes your writing clearer or more concise, use it.

The chemist poured water into an open glass cylinder with a pouring lip.	The chemist poured water into a beaker.

But avoid technical terms when a simpler word will do just as well, or when the term's meaning may be unclear to a significant portion of your readers.

The moon was in syzygy.	The moon was aligned with the sun and the earth.
Maximize the decibel level.	Turn up the sound.
Stabilize mobile dentition	Keep loose teeth in place

Rule 43. Break up the writing into short sections.

In technical writing, it is not uncommon to find sentences of more than 40 words, or paragraphs that run on for pages. While some ideas may truly need this type of elaboration, a great many do not.

Since readers have a built-in fondness for brevity, you should constantly look for ways in which to break long sentences and paragraphs into shorter, easier-to-grasp units. Surely, in any 50- or 60-word sentence, there's at least one opportunity to divide the sentence into two shorter sentences. And a paragraph of 10 to 15 sentences might be broken into two or more shorter paragraphs by finding places where a new thought or idea is introduced and beginning the new paragraph with that thought.

In the same way, the text itself should be broken up into many short sections and subsections. Use headlines and subheadlines to title each section. Each short section should be a self-contained mini-essay on a single topic or thought. Experience teaches us that a piece of writing is most effective when it deals with *one simple idea*.

Rule 44. Keep ideas in writing parallel.

Parallel sentence structure exists when two or more sentence elements of equal importance are similarly expressed. The benefits of parallelism are many: an economy of words, a clarification of meaning, a sense of symmetry, and a sense of the equality of each idea in the sentence. In fact, the previous sentence is a good example of parallelism.

Here are a few other examples:

> The tube runs into the chest cavity, across the lungs, and into the stomach.

> The atomic weight of gold is 196.97; silver, 107.87; iron, 55.85; lead, 207.19.

> ''Ask not what your country can do for you, ask what you can do for your country.''

> ''It was the best of times and the worst of times. . .''

Rule 45. Opt for an informal rather than a formal style.

Do not hide behind an overly formal writing style. We are not recommending that your technical papers be written in the same chummy tone as a letter to your pen pal. But a more relaxed, conversational style *can* add conviction and vigor to your work.

Technical writers especially go to great lengths to avoid using personal pronouns. This can result in an unnaturally stiff style. Why conceal your identity? Instead of writing *the measurement was taken*, write *I took the measurement*—if, indeed, you did take the measurement. Here are some examples of formal versus informal style:

Formal	*Informal*
It is unfortunate that I was not available when you visited our facilities the other day.	I'm sorry I missed you the other day.

For the purpose of breaking up a beam of sunlight into the seven visible colors of the spectrum, a glass prism was procured.	I obtained a glass prism to break up sunlight into a rainbow.
A system can be designed to meet your requirements.	We can design a system to meet your needs.

In particular, avoid dated transition words such as *whereby, heretofore, herein,* and *wherein.* If your writing reads as if you went to law school, you have fallen under the spell of "legalese" (also known as "bureaucratese," "federalese," or "corporitis") in which, for no apparent reason, words take on a certain institutional stiffness. When words such as *wherein, thereby, heretofore,* and *whereby* creep into your vocabulary, put down your pen, take a few deep breaths, and read your work aloud. Your ear will soon tell you just how awkward and antiquated these phrases are.

perchance	This sounds like something Shakespeare might have said almost 400 years ago ("to sleep . . . perchance to dream"); it is a bit poetic for technical writing.
persuant to my request	This is another weighty phrase that just fills up space.
hitherto	Why not simply use *until now?*
enclosed herewith	Herewith? *Where*with? It sounds pretentious as well as starchy.
as per	This is undesirable jargon for *in accordance with* or *as you suggested.*
inasmuch	Archaic
attached hereto	Stilted, awkward

enclosed herein	Stilted
whereof	Stilted
thereof	Stilted
thereby	Stilted
thereto	Thereto? *Where*to?
whereby	Archaic
of even date	"Insurancese" for *today*. Why not just write *today?*
persuant to your orders	Overly formal. Just write *following your directions.*
keep me timely advised	Insurancese for *let me know as soon as you do*—a phrase which at least tells laypeople what is going on!
aforementioned	Unless you write leases, avoid this lawyerlike expression.
whereas	Instead use *where* or *while.*
etc.	Instead of *A, B, C, etc.,* write *A, B, and C. Etc.* has its uses, but in general contexts it is a lazy way out of a problem. Avoid it.

6

Words and Phrases Commonly Misused in Technical Writing

TECHNICAL WORDS AND JARGON

Every field has its own technical vocabulary: a special language that helps specialists describe a concept, process, or thing. The term *standard deviation* has a precise meaning to statisticians, in the same way that *excess* has special meaning to an insurance claims adjuster.

Technical words such as these are helpful and necessary. Problems arise, however, when technical words proliferate, and some become a slanglike shorthand for precise communication. At that point, technical terms turn into "buzzwords," and these catch phrases hoodwink writers, making them forget that the words have limited application and may be understood by only a few insiders.

Don't throw around jargon. A hospital administrator might use the term *catchment area* to describe the place from which hospital patients may be drawn. But this term sounds strange to people who are not in the health services and may cause an outsider to be confused (and possibly giggle).

In the same way, how much is the technical writer communicating to the nontechnical reader when, in an automobile advertisement in *Time*, she or he writes that a particular model has an *electronic instrument cluster, an electronic bar chart fuel gauge, an optimum 3.8 liter V-6 engine, an automatic overdrive trans-*

*mission, a modified MacPherson strut front, supervision and sta-
bilizer bars, and 4-bar link rear suspension?*

Even everyday words can have many meanings to people in different specialties. To doctors, a tongue is what they tell patients
to stick out of their mouths. But concert organists know the tongue
as vibrating slips of metals producing the tones in their instruments.
Likewise, if you say *drone,* biologists think of a male honeybee,
military personnel of remote-control aircraft, and Scotsmen of the
continuous-sounding pipes in their bagpipes.

*Technical writers need to decide when they are using appro-
priate terms and when they are obscuring their meaning in needless
slang or bombarding the reader with technical overkill.*

Do not invent technical words just to add a touch of importance
to your writing. We once saw a memorandum dealing with park
"signage," and it took us several minutes to figure out that the
subject was *signs.* Inventing and perpetuating terms such as *on-
line, outage, stagflation, supply-side economics, grid-lock, feed-
back,* and *cost-effective* may be fun—especially for the media—
but the careless writer may forget that these words are somewhat
less than universal. If two computers are not *interfacing,* it may
be a serious problem; but when a writer states that two executives
don't *interface,* we worry more about the creator of the phrase than
about those poor noninterfacing executives!

When words or phrases such as *meltdown, half-life,* and *burnout*
become ingrained in a writer's vocabulary, they begin the long
slow slide into clichédom and oblivion. While they are on their
way, these words tend to alienate readers from the precise, unsim-
plified meanings they only hint at.

Once you get over the temptation to make your writing sound
important, to be a verbal show-off, you will have taken a big first
step toward simplicity and clarity. A medical advertisement that
we saw recently serves as a good example. It described a drug
which shows a "low incidence of adverse reactions." Translated
this means that the drug has few side effects.

BIG WORDS

Technical writers sometimes prefer big, important-sounding
words to shorter, plainer language. As the syllables grow, the writ-

ing may sound more "important," but it becomes harder to understand. When sentences get long and words balloon in size, the mind of even the most patient and intelligent reader is likely to wander.

If the technical writers could achieve some aesthetic distance from their writing and put themselves in the reader's place, they would search for simple words instead of long, complicated ones to express their thoughts. Too often, technical writers forget that their audience may include people outside the writers' department, industry, or discipline.

In electronics, for example, technicians do not *free* a link; they *disengage* it. Laboratory experiments are never *ended,* they are *terminated*. And, of course, engineers never *estimate;* they *approximate*. (Even airplane personnel prefer the cumbersome term *deplane* to *leave* or *get off*.)

Do not use a big word when a smaller one will do.

Here are a few big words in common use. The column on the right shows a simpler—and preferable—substitution:

Instead of	*Use*
aggregate	total, whole
amorphous	shapeless
anomalous	abnormal
antithesis	opposite
abbreviate	shorten
ascertain	find out whether
aqueous	watery
autonomous	independent
beverage	drink
circuitous	roundabout
commencement	start, beginning
contiguous	near, touching
comestibles	food
coagulation	clotting, thickening
cessation	stop, pause
conjecture	guess
concept	idea
currently	now

Instead of	*Use*
discourse	talk
disengage	free
demonstrate	show
deficit	shortage
duplicate	copy
expedite	hasten, speed
elucidate	clarify
eliminate	cut out
facilitate	ease, simplify
feasible	possible
gradient	slope
homogeneous	uniform, similar
incombustible	fireproof
incision	cut
impairment	injury, harm
inundate	flood
miniscule	tiny
maintenance	upkeep
nomenclature	name, system of terms
orientate	orient
obtain	get
optimum	best
posterior	rear
parameter	variable, factor
potentiality	potential
requisite	needed, necessary
subsequent	next
sufficient	enough
segregate	set apart
terminate	end
verification	proof
viable	workable
vitreous	glassy

WORDY PHRASES

Avoid the wordy phrase; strive to be succinct. During a 1981 World Series game, sportscaster Howard Cosell commented that

Yankee owner George Steinbrenner and manager Billy Martin had "a mutuality of affection for each other." We suppose that means they *like* each other.

When you edit your writing, simplify those wordy phrases that take up space but add little to meaning or clarity. The following list includes some common wordy phrases. The column on the right offers suggested substitutes.

Wordy Phrase	*Substitute*
despite the fact that	although, even though
at this point in time	at this time, now
during the course of	during
on an annual basis	yearly
in the vicinity of	near
on a weekly basis	weekly
on the occasion of	when
in the majority of instances	usually, most often
in the form of	as
until such time as	until
take action	act
on the basis of	by, from
hold a meeting	meet
in the process of tabulating	in tabulating
have discussion of	discuss
subsequent to	after
the reason why is that	because
with reference to	about
with the result that	so that
in the event that, of	if
prior to that time	before
has proved itself to be	has proved, is
in the form of	as
in order to	to
in many cases	often
in some cases, in other cases	sometimes
in most cases	usually, often
as shown in table 6	table 6 shows

Wordy Phrase	*Substitute*
has been widely acknowledged to be	is
the necessity is eliminated	you do not need to
exhibits the ability	can
could be considered as	is
is reported to be	is
at the time of presenting this paper	today
even more significant	more significant
it is clear that	clearly
is equipped with	has, contains
in the course of	during, while
a large number of	many

Some phrases are so inflated and meaningless that they can be omitted entirely. They merely take up space and should be deleted from every sentence.

> the fact that
> it has been shown that
> it is recognized that
> it has been demonstrated that
> has gained much more importance in recent years
> it must be remembered that
> it may be seen that
> what is known as
> it is worthy of note
> it will be appreciated that
> it is found that
> it may be mentioned that
> it is the intention of this writer to
> thanking you in advance for your cooperation
> deemed it necessary to

REDUNDANCIES

Redundancies are an insidious form of wordiness which inflict themselves on the work of nontechnical as well as technical writers. Redundant words announce their own superfluity because typically a modifier merely repeats an idea already contained in the word being modified.

Notice how many people use the phrase *very unique*. Actually, *unique* means one of a kind, so it is impossible for anything to be *very* unique.

One car manufacturer designed its advertising campaign around the slogan *new innovations*. Could there be such a thing as an *old* innovation? The best way to spot a redundancy is to ask what a word is "buying" in a particular phrase. Is it *adding* meaning? Or is the word there because the writer was not sure that the thought had been communicated completely by using *unique* or *innovation* or any other sharply defined term?

Consider the phrase *study in depth*. Ask yourself: Doesn't the word *study* already imply *in depth*? In the same way, *consensus of opinion* is a redundancy because *consensus*, by itself, implies a solidarity of group opinion.

When redundancies are looked at in this way, writers can spot them easily. A phrase such as *investigative reporter* suddenly sounds odd, because *all* reporters investigate. Could there be such a thing as a *non*investigative reporter? Here's a list of other common redundancies, along with suggested substitutions:

Redundancy	*Substitution*
advance plan	plan
an honor and a privilege	an honor
any and all	any
absolutely essential	essential
absolutely perfect	perfect
continue on	continue
consecutive in a row	consecutive
basic essentials	essentials
close proximity	close
first priority	priority
first introduction	introduction
combine into one	combine
current status	status
equally as well	equally
final outcome	outcome
goals and objectives	goals

Redundancy	*Substitution*
honest truth	truth
joined together	joined
one and the same	the same
isolated by himself	isolated
overall plan	plan
point in time	time
personal opinion	opinion
first and foremost	first
repeat again	repeat
refer back to	refer to
true facts	facts
this particular instance	this instance
different varieties	varieties
wrote away for	wrote for
take action	act
you may or may not know	you may know
whether or not	whether
all of	all
by means of	by
adding together	adding
small in size	small
past history	history
necessary requisite	requisite
as a general rule	as a rule
connected with	with, in, of
uniformly consistent	consistent
16 cubic meters in volume	16 cubic meters
triangular in shape	triangular
actual experience	experience
balance against one another	balance
cancel out	cancel
range all the way from	range from

CLICHÉS

Clichés are words or phrases that have worn out their welcome—
in fact, *worn out their welcome* is a good example of a cliché! They
have become trite through overuse, and so they no longer com-

municate in the same crisp way they once did. In a sense, they have been penalized for their popularity. Here are a few clichés which should be used sparingly if at all:

acid test	the light of day
bottom line	point in time
ballpark	state of the art
blaze	hands-on
beyond the shadow of a doubt	feedback
back to square one	meaningful
cost-effective	overkill
dealing with	try it on for size
dialogue	top dollar
bread-and-butter issue	under review
coping	richly deserved
escalate	vitally important
eyeball (used as a verb)	tried and true
great success	viable
grind to a halt	relevant
a no-no	

OVERBLOWN PHRASES

Words reflect society, and when society changes, so do its words. However, some phrases seem to linger. Perhaps these words have become so comfortable that we perpetuate them instead of retiring them.

You can spot an overblown phrase—whether it's antiquated, pompous, or just a meaningless stock phrase—by reading your writing aloud. When we read aloud the following phrases, we are reminded just how ludicrous or patronizing they have become.

I'm sure you can appreciate	This phrase is patronizing and should be avoided.
note how this matter will be handled	Also patronizing.
when time permits	It may be poetic, but it's also inaccurate. Time doesn't permit; people do.

by virtue of	To paraphrase the late Mae West, "virtue has nothing to do with it." *Because* is usually an effective substitute.
kindly advise	As opposed to *unkindly?* Unnecessary.
don't hesitate to	What is really meant is *Feel free to*.
deemed it necessary	Old-fashioned.
under separate cover	Also old-fashioned.

THE RISE OF *IZE*

Where did it begin, this rush to add *ize* to words? Why do we feel compelled to create verbs where no verb existed? Edwin Newman, in his book *A Civil Tongue,* suggests that by adding *ize* to certain words, people believe they are adding a businesslike tone. *Prioritize* sounds more businesslike than *make priorities,* but it is also awkward, pretentious, and incorrect. *Utilize* may sound more technical than *use*, but it adds only length, not meaning. Also, there's the problem of ambiguity: does *finalize* mean to complete? And just what is being completed? By using *finalize,* writers slip into vagueness by failing to tell us precisely what has been concluded. It could be the signing of a document, the agreement to wording, the agreement to even write a contract. Therefore, the finalizing of a document is too vague and general to really pass along meaning.

Here are a few other *ize* words to be used sparingly, if at all:

maximize	academize
minimize	normalize
customize	strategize
standardize	definitize
optimize	traumatize
politicize	formalize

NOUNS AS ADJECTIVES

Strictly speaking, it is bad grammar to use a noun as an adjective. But today it's common practice; using nouns as adjectives can make a sentence more direct, less wordy. For example, the phrase *an iron pipe* uses the noun *iron* as an adjective modifying *pipe*. Yet, it's better than the more roundabout phrase *a pipe made of iron*.

Nouns as adjectives are acceptable when used *sparingly*. In technical writing, the trouble starts when writers string nouns together to make cumbersome, hard-to-swallow phrases.

We have all had the experience of watching adjectives proliferate until they actually obscure, instead of illuminate, meaning. For example, if you were describing a power boiler at an industrial plant, you might write

an industrial power boiler

Let's say you had to include the power rating and fuel. Now you write

a 15,000 lb steam/h pulverized coal-fired industrial power boiler

If too many adjectives pile up, break up the list with prepositional phrases.

a 15,000 lb steam/h pulverized coal-fired power boiler	a pulverized coal-fired power boiler generating 15,000 pounds of steam per hour
chronic post-traumatic stress disorder	a chronic stress disorder following trauma
a frequency-shift, power-line carrier relaying system	a relaying system using a frequency-shift power-line carrier
software-programmable modular information retrieval system	a modular information retrieval system with programmable software

| long-range, high-power, three-dimensional surface radar | a high-power surface radar that detects long-range targets in three dimensions (range, azimuth, and elevation) |

MISUSED AND TROUBLESOME WORDS AND PHRASES

English is a rich language, and so it is inevitable that words with subtle shades of meaning are often confused and misused. Some are misused because slang has made their meaning muddy; others because their actual definition is overshadowed by what the public believes their definition to be; still others because they sound alike and are often mistaken for one another.

The following is a selected list of commonly misused words and phrases. Since the technical writer generally must uphold a higher standard of accuracy than the nontechnical writer, you may wish to review these words by reading them aloud, studying their definitions, and using them in sentences.

ability, capacity	*Ability* means the state of being able or the power to do something. *Capacity* is the power of receiving or containing.
about, approximately	*About* indicates a guess or rough estimate (about half full). *Approximately* implies accuracy (approximately 34.76 gallons).
accept, except	*Accept* means to receive willingly, to agree with. *Except* means excluding.
advise, inform	*Advise* means to offer counsel and suggestions (I advise you to sell that stock). *Inform* means to communicate information (I inform you that your shipment has not arrived yet).
affect, effect	*Affect* is a verb meaning to

	change or influence. *Effect* is a noun meaning result or outcome. *Effect* is also a verb meaning to bring about.
aggravate	*Aggravate* means to make worse. Don't use it as a synonym for *irritate, annoy,* or *provoke*.
all together, altogether	The phrase *all together* means that everyone is in the same location. *Altogether* means entirely.
allude, refer	To *allude* is to refer indirectly. To *refer* to is to name.
and/or	This is an awkward construction. Avoid it.
anxious, eager	Use *anxious* when anxiety or worry is involved, not as a synonym for eager. *Eager* means highly desirous of something.
because of, due to	*Because of* means by reason of or on account of. (The radar failed because of a short circuit.) *Due to* means attributable to. (Due to his efforts, the radar was repaired in an hour.)
beside, besides	*Beside* means by the side of. *Besides* means in addition to.
between, among	Use *between* when writing of two things. Use *among* when writing of three or more.
big, large, great	*Big* is used to refer to bulk, mass, weight, or volume. *Large* is used with nouns indicating dimensions, extent, quantity, or capacity. *Great* is now used almost entirely to connote importance or eminence.

can, may	*Can* implies ability; *may* implies permission.
center, middle	These terms are not interchangeable. From its geometric definition, *center* retains, even in nontechnical contexts, the idea of a point around which everything else revolves or rotates. *Middle* is less precise, suggesting a space rather than a point.
continual, continuous	*Continual* means recurring frequently; *continuous* means without interruption.
data, datum	When *data* is used synonymously with *facts,* it is plural. When it is used synonymously with *information,* it is singular. The singular form *datum* has fallen out of popular use in technical writing.
disinterested, uninterested	*Disinterested* means impartial. *Uninterested* means indifferent, having no pleasure or delight in something.
effective, efficient	A machine that's *effective* performs its intended function well. If it does this with a minimum of waste, expense, and unnecessary effort, then it's *efficient* as well.
e.g., i.e.	*e.g.* means for example; *i.e.* means in other words or that is.
equipment, equipments	*Equipment* is both singular and plural. There is no such word as *equipments.*
ensure, insure	*Ensure* means to make sure of

something. *Insure* means to take out an insurance policy.

everyone, every one

Everyone means all people. *Every one* means each one.

farther, further

Farther refers to physical distance. (Pluto is *farther* from the sun than Earth.) *Further* refers to matters in which physical measurement is impossible. (*Further* research would be useful.)

fewer, less

Fewer is used when units or individuals can be counted (*fewer* containers). *Less* is used with quantities of mass, bulk, or volume (*less* weight).

hopefully

The phrase *hopefully the situation will improve* is ridiculous because the situation cannot be full of hope. *Hopefully we shall fly to Pittsburgh tomorrow* does not mean we can hope to be in Pittsburgh tomorrow. It means we shall fly there full of hope. Beware *hopefully*.

impact

Impact does *not* mean to affect or influence. *To impact* means to drive or press closely into something. Avoid *impact* as a synonym for *effect*.

imply, infer

Imply means to signify or suggest something. *Infer* means to deduce or conclude something. (After studying our data, Bill *inferred* that product degradation had occurred. He *implied* that a faulty heat exchanger might be the result.)

incredible, incredulous

Incredible means not believable. *Incredulous* means not able to believe. Statements or assertions are *incredible*; people are *incredulous*.

electric, electrical

Electric refers to anything that produces, carries, or is activated by electric current (electric appliance, electric car, electric circuit). *Electrical* refers to anything that pertains to but does not carry electricity (electrical engineer, electrical insulation).

irregardless

This is not a word. Use *regardless*.

like, as

Like is still not accepted as a conjunction except when it introduces a noun not followed by a verb. (He builds bridges *like* a veteran architect).

materiel, material

Materiel is the equipment, apparatus, and supplies used by an organization. *Material* refers to the substances of which something is composed.

noted, notorious

People famous in a desirable way are *noted;* people with unsavory reputations are *notorious*.

over, more than

Over implies position. Do not write *over* when you mean more than. (There are *more than* 100 installations worldwide. I would choose this system *over* any of the others).

practicable, practical

Practicable means that which appears to be capable of being put into practice. *Practical* means that something is known to be doable based on past performance.

presently, at present

Presently means soon; *at present* means now.

principal, principle

Principal used as a noun means head of a school, a main participant, a sum of money. As an adjective, it means first or highest in rank, worth, or importance. A *principle* is a fundamental law, a basic truth.

should, will

Should implies ought to, a belief. *Will* is a prediction.

sic

Sic is Latin for *thus,* and it is used in quotations to indicate that the writer has quoted the material exactly as it appears in the original source.

that, which

Ideally, *that* is used with a restrictive clause—a clause absolutely necessary to the sentence. (This is the vessel *that* holds the acid.)

Which is used with a nonrestrictive clause—a clause that adds descriptive matter and is not necessary to the sentence. (The steel vessel, *which* is used to hold acid, was lined with ceramic bricks.)

ultimate, penultimate

Ultimate means last. *Penultimate* means next to last. Do not use *penultimate* as a superlative of *ultimate*.

unique

No superlatives are needed, since *unique* means one of a kind. Therefore, *really unique, so unique,* and similar slang are grammatically incorrect.

In this section, we covered the use of less than 100 words. Your writing vocabulary may be as large as 20,000 words or more, so not all your questions about proper word choice will be answered by the examples in this book.

Before you use a word, ask yourself three questions:

1. Is there a simpler, smaller word that would get my meaning across just as well?
2. Will the majority of my readers understand the word?
3. Is the word as specific and concrete as it can be?

If the word is too complex, too big, unfamiliar, vague, or abstract, search for another word that will serve you better.

7

Producing the Technical Document

Unlike novelists, newspaper reporters, and magazine journalists, technical writers do more than write: They have to turn manuscripts into polished, printed documents. This means that technical writers must edit the text as well as coordinate typing, illustrations, production, printing, and distribution. Technical writers, then, are also *producers*.

Because of the highly technical nature of the material they work with, technical writers encounter special problems in production. In this chapter, we see how technical editing, technical typing, and technical illustrations differ from their counterparts in nontechnical publishing.

TECHNICAL EDITING

What's the difference between a *technical writer* and a *technical editor?* And what, exactly, does an editor do?

Although job descriptions are often vague, a *technical writer,* strictly speaking, is the *author* of the document: The technical writer writes the first draft. With highly technical material—reports, proposals, and papers—the scientist or engineer who is closest to the subject is in the best position to do the writing. With brochures, manuals, advertising, and other less technical material, a professional technical writer may be the author.

Technical professionals have little time to fuss with the details of editing, typing, and printing, and so they hand over their first drafts to the *technical editor* for polishing and production. According to *The American Heritage Dictionary,* it's the editor's job "to make (written material) suitable for publication or presentation." Ruth M. Power, a technical editor at Badger America, Inc., says the technical editor's job "is one of quality control for written communications."[1]

Earlier we stated that the nontechnical writer is concerned with literary style, while the technical writer is more interested in the effective communication of technical information. The same goes for editors. The nontechnical editor working for a magazine, newspaper, or book publisher shapes a manuscript into a good piece of writing—something with imagination, flair, vividness, and style. This literary editor's chief concerns are:

Good use of the language

Literary style—pace, flow, transition, and tone

Grammar, punctuation, capitalization, and spelling

Sentence and paragraph structure

Consistency in format and style

The technical editor checks for all these things, too. But, her or his main interest is communicating technical information, not creating a literary masterpiece. More than style and language, the technical editor checks for the following:

Technical content—completeness, accuracy, and conciseness

Compatibility of the technical depth of the text and the reader's background

Effective communication of technical information in a clear, logical, sensible manner

Proper use of technical terms and symbols

Effective coordination of text and artwork

[1] Ruth M. Power, "Who Needs a Technical Editor?" *Hydrocarbon Processing,* February 1981, pp. 167–171.

Many technical professionals think of editors as technically unknowledgeable "liberal-arts types" who cross out sentences at random just to make documents less bulky. Nothing could be further from the truth. It's the technical editor's job to help the reader by making publications easier, less time-consuming, and more enjoyable to read. Editors do this by editing text to make it concise, clear, and precise. Editing is really four separate tasks: proofreading, deleting, repositioning, and rewriting.

Proofreading

Proofreading is standardization. Proofreaders make writing consistent in the use of capitalization, abbreviations, and grammar. They correct spelling and punctuation errors. And they check typed text against the original manuscript. Proofreaders check the picayune details that many authors ignore but find so important when the document is published.

Deleting

A large part of the editor's job is to strike out all unnecessary words, sentences, paragraphs, pages, and illustrations. By eliminating extraneous technical material, the editor makes the document less confusing and cuts down on reading time.

Repositioning

The right word order can make your writing flow. But it's hard to write precisely when you are "getting it all down on paper" in your first draft. The technical editor repositions words to make your meaning clearer and rearranges paragraphs and pages to make the organization of the document more logical.

Rewriting

At times, an editor will come across text so jumbled that it will have to be rewritten entirely. Here the editor must work closely with the author to find out what was really intended.

If you edit other people's work, do not rewrite unless it is necessary. Remember, it is the writer's job to write, the editor's to edit.

TECHNICAL TYPING

Technical typing is more difficult than everyday office work because Greek letters and scientific symbols cannot be typed on many standard typewriters. Also, accurately typing equations and tables is a tricky job that takes practice—and patience.

To achieve technical accuracy and a professional appearance in your technical typing, first you need the right equipment. If you are going to reproduce documents directly from typed manuscripts, you will need a good electric typewriter, preferably one with interchangeable keys or elements and a self-erasing feature. IBM's Selectric and Composer typewriters have interchangeable elements ("golf balls") containing Greek letters and scientific, technical, and mathematical symbols.

For other typewriters, you can use Typit. This device fits on your machine and lets you quickly convert the keys from technical type to regular and back again.

At art supply stores, you can purchase pressure-sensitive (rub-on) lettering, transfer lettering, or templates that let you make consistent, professional-looking technical symbols by hand. Some writers try to save time by writing equations and symbols freehand, but this looks amateurish and reproduces poorly.

Finally, organizations that produce a large volume of technical documents might consider investing in word processors. These machines can type, edit, and store text electronically. This is especially useful for large documents that must be updated frequently, such as manuals.

In typing technical documents, care should be taken to avoid confusion between symbols and standard letters. For example, the capital letter X looks like the multiplication sign, the small letter l like the numeral one, the capital letter O like zero.

Mathematical equations and expressions must be typed in one consistent style. There is no standard spacing; typists should find the style that looks best on their machines and stick with it. We offer some guidelines for technical typing, but these are only suggestions, not strict rules.

Some guidelines for the technical typist

1. Superscripts are typed one-half space above the character they modify; subscripts, one-half space below.

$$x^2$$
$$F_s$$

2. When a superscript is a simple fraction, use the solidus (/) instead of the horizontal line.

$$T^{1/2} \quad \textit{not} \quad T^{\frac{1}{2}}$$

3. A space precedes and follows operational signs ($=$, $+$, $-$, $\pm$, $>$, $\geq$, $<$, $\leq$) in the main line of an equation, underneath the square root sign, or in the numerator or denominator of a fraction. This is known as *open spacing*.

$$y = mx + b$$
$$a = \sqrt{b^2 + c^2}$$
$$r = \frac{t + v}{p - q}$$

However, it is not uncommon to omit the spaces around the equals sign.

$$y = mx + b$$

When plus and minus signs indicate the positive and negative values of a term, and not the mathematical operations of addition or subtraction, no space follows the plus or minus sign.

$$T = -44°C$$

4. Leave large margins. Technical editors need extra room to write instructions for typesetters who have to deal with complicated equations and mathematical expressions. We suggest margins of at least 1½ inches.

5. Double-space documents that will be edited and marked with instructions for the typesetter. It is acceptable to use single space if the document will be reproduced directly from the typed manuscript page. Use good-quality, standard 8½ by 11 inch white bond typing paper.

With a little extra care and effort, you can produce first-class technical manuscripts on your office typewriter. Often, you will have to illustrate these texts before you can reproduce the work.

TECHNICAL ILLUSTRATIONS

In books, magazines, and advertising, photographs and illustrations can be used to make the publication more pleasing to the eye and to gain the reader's attention.

In technical publications, however, visuals are functional, not ornamental. They are a means of presenting complex information and ideas. Studies show that visuals can communicate more effectively than words: Average readers remember only 10 percent of what they read, but 30 percent of what they see.[2]

Technical writing is illustrated with a wide variety of visuals. A schematic diagram can show the electric wiring of a transmitter or the layout of an airplane interior. Cutaway drawings let readers see the interior construction of equipment and systems. A photograph can offer tangible proof of claims made by the writer. A graph can track how air temperature changes with the time of year, or how your blood pressure increases with physical activity. Photographs, drawings, maps, diagrams, graphs, charts, and tables are the basic tools of the technical illustrator, and we briefly outlined their use in Chapter 2 (see the table on page 23).

How do technical writers get their work illustrated? The cheapest method is to draw the graphs and take the photographs yourself. However, the problem with the do-it-yourself approach is that it is apt to *look* as though you did it yourself.

Many organizations have people who can do neat, reproducible line art—lettering, schematics, graphs, and simple line drawings. Perhaps there is even a photographer or technical illustrator in the house. Look around, and take full advantage of your internal resources. If no one in your organization can handle the job, numerous free-lancers are available who can give you top-notch work.

Unless you are producing advertising or promotional material, avoid elaborate artwork. Use simple, uncluttered illustrations—

[2]Edward Palmer, ''Audio Visuals in Corporate Communications—70 Years of Success,'' *Audio Visual Directions*, May/June 1981, p. 76.

visuals that are attractive, readable, and interesting to look at. If you are reproducing the document on your office copier or at the corner print shop, the graphs and charts should be solid black ink on white paper or art board. Tell the artist what you need and how it will be reproduced, and you will be provided with a *camera-ready mechanical*—artwork prepared in a form acceptable to the printer.

There are no universal rules for the proper use of visuals in technical publications. Below, we suggest some guidelines based on experience and common sense.

How to illustrate technical publications

Title, caption, and number visuals. Every visual should have a caption or title. This label should be self-explanatory; your reader should not have to refer to the text to understand the illustration.

Jane Maas, a leading authority on sales brochures, estimates that the readership of photo captions is almost double that of body copy.[3] Therefore, it pays to write informative, detailed captions. Do not label a graph, *World energy consumption,* when you can write, *World energy consumption has experienced an average growth rate of 5 percent a year for the past 10 years.*

It is not mandatory that you number every visual, but it is useful when you want to refer to a table, an illustration, or a graph in the text. Tables are numbered Table 1, Table 2, Table 3, and all other visuals (photographs, drawings, graphs, charts) are called figures (abbreviated "Fig.").

Reference visuals. There are several ways of referring the reader to a figure or table.

> As can be seen in Fig. 4. . . .
> Figure 2 shows. . . .
> This system (see Fig. 12). . . .
> Table 1 presents the data . . . .

[3]Jane Maas, *Better Brochures, Catalogs and Mailing Pieces,* St. Martin's, New York, 1981, p. 19.

Note: Abbreviations, such as Fig., are not acceptable at the beginning of a sentence; use the spelled-out form or rewrite.

Avoid graphic clutter. Technical professionals tend to cram every last detail into their tables and illustrations. But too much information can clutter a visual, making it sloppy and hard to read.

If you must present a great deal of information, make your visual large. (You can even use a fold-out page larger than the standard 8½ by 11 inches.) Or else use more than one visual to tell your story. But crowding figures with too many small numbers and lines will only frustrate, not help, your readers.

Information in visuals should be consistent with symbols and units used in the text. One way to cut production costs is to borrow existing artwork and use it in your publication. Other people in your organization may have already paid artists to produce visuals that are applicable to your own subject.

Production and advertising people call this recycling of material *merchandising*. It will stretch your budget, but let the technical writer beware: The symbols, units, and technical terms in these borrowed figures may not be consistent with what is used in the writing. This inconsistency can muddle your meaning and confuse the reader. Fortunately, there are two simple solutions to the problem. The first is to change the visual to conform with your choice of nomenclature and units. This may involve some expensive artist's alterations. A simpler, less costly solution is to include a key with the figure, showing the relationship between its symbols and yours.

Place each visual on a separate page following the page of the first mention. The pages on which visuals appear are numbered in sequence with the rest of the pages. (Some editors prefer to combine text and illustrations on the same page.)

Never have an illustration that presents itself upside down when the document is held in the normal reading position or with the "gutter" (the white space between facing pages) at the top.

(a) The best position for a visual

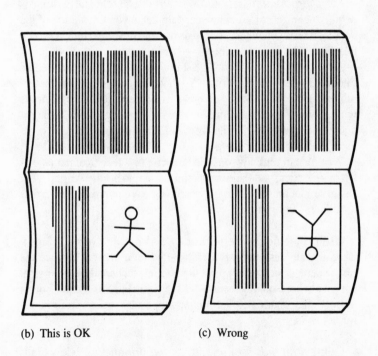

(b) This is OK   (c) Wrong

Normally, a figure should present itself right side up when the document is held in the normal reading position (a); if it does not fit this way, it should present itself right side up when the bound edge is at the top (b).

GETTING IT PRODUCED AND PRINTED

Any professional technical editor can tell you that coordinating production and printing requires careful planning. Missed deadlines, apathetic authors, last-minute changes, and other problems can make it difficult to publish good work on time.

We have a few helpful hints for harried technical editors and production managers. These tips will not solve every problem, but they will make production go more smoothly and help you meet those ''I need it yesterday'' deadlines.

Production tips for technical editors

Put a production schedule in writing. Technical writing is detailed work. A production schedule helps you keep track of the details. It tells you what needs to be done, when, how long it should take, and who is responsible.

As a technical editor, you must rely on input from other people in order to get your job done. A schedule spells out the responsibilities of the engineers, scientists, managers, and other authors who are writing sections, providing data, drawing graphs, and producing the raw material that you will mold into a cohesive, organized document.

Writing can rank low on an engineer's or a technical manager's list of priorities, and their late submissions and hastily written work can give a technical editor gray hairs. Give your authors and artists deadlines—*in writing*.

Use expert help when you can afford it. Advertising agencies, graphic designers, professional editors, and printers are the real experts at producing printed material, technical or otherwise. On jobs where quality counts—advertisements, brochures, other promotional material, even major proposals—ask a professional for help.

Get several bids for production and printing. The low bid is not necessarily the best bid—the creativity and professionalism of your outside suppliers are important, too. But competitive bids ensure that a supplier's price is not out of line.

Use original artwork and photographs. Photographs clipped from trade publications and diagrams run off on the office copier will reproduce poorly. Try to print from original mechanicals and photographic negatives or first-generation transparencies.

Obtain all necessary approvals in writing before production is begun. In most companies, all external communications—advertisements, brochures, press releases, and technical articles—must be approved by several layers of management before they can be released outside the corporation. Attach an approval sheet, circulate your typed manuscript to the people who must approve it, and have them sign their approval before passing the document to the next reviewer. This will save you money, since it is easier and less expensive to change a typed manuscript than an artist's mechanical or a printer's plate. And, frankly, you won't take the blame for other people's mistakes and changes if you have a signed, approved copy sitting in your files.

Proofread every step of the way. Your readers will think less of you, your organization, and your work if your writing is full of typos, misspellings, and other errors. Proofread thoroughly every step of the way—the typed manuscript, the mechanicals, the printer's "blue-line" (a single copy of the document made from the plates or films just before the copies are printed).

Have a number of people proofread the mechanicals. Sometimes editors find mistakes even after 10 proofreadings.

GETTING IT DISTRIBUTED, PRESENTED, OR PUBLISHED

When the printer delivers your printed documents, it is up to the technical writer to see that they are distributed. Make sure everybody who might find the publication useful is on the distribution list. The list may include customers, sales representatives, field offices, and your boss, as well as the engineering, production, sales, marketing, advertising, manufacturing, technical publications, and research departments.

Brochures, bulletins, catalogs, reports, and proposals are "self-published" in that you pay a printer to reproduce them. Articles and papers, however, are published in trade or scientific journals or presented at technical conferences. (The journal editor or conference chairperson usually will supply reprints for a reasonable fee.) If you want to publish or present a technical paper, select a magazine or meeting where your work will reach a large and interested audience. *Bacon's Publicity Checker* and *Writer's Market*, two books available in your local library, list trade and scientific publications by specialty. (Under *Bacon's* listing for electronic engineering, for example, you will find such journals as *Computer Design, Electronic Buyer's News, Electronic Warfare Design,* and *Holosphere*.) In fact, the journals in your field that you already read are probably the best place to publish your paper.

Once you have selected a publication, contact the editors to see whether they are interested in your proposed article. If so, they will ask for a manuscript and set a deadline. Be sure to obtain the magazine's guidelines for authors, so that the format and style of your manuscript fit their editorial requirements.

Most magazines carry listings of technical meetings. If you intend to present a paper, you will be required to submit an *abstract* (see Chapter 2) to the meeting chairperson. With a relevant topic and a well-written abstract you can count on being invited to the conference—as a speaker.

APPENDIXES

Greek Alphabet

Name	Typed		Handwritten Version
	Uppercase	Lowercase	
Alpha	A	α	α α
Beta	B	β	β
Gamma	Γ	γ	γ
Delta	Δ	δ, ∂	δ ∂
Epsilon	E	ε	$\mathcal{E}$ ϵ
Zeta	Z	ζ	η ς s h h
Eta	H	η	η
Theta	Θ	θ	θ θ φ
Iota	I	ι	ι
Kappa	K	κ	κ κ
Lambda	Λ	λ	λ λ
Mu	M	μ	μ $\mathcal{U}$
Nu	N	ν	υ $\supset$
Xi	Ξ	ξ	ξ γ
Omicron	O	o	o
Pi	Π	π	π
Rho	P	ρ	ρ ρ ℓ
Sigma	Σ	σ	6 σ
Tau	T	τ	τ
Upsilon	Υ	υ	γ
Phi	Φ	ϕ, φ	ϕ φ
Chi	X	χ	χ
Psi	Ψ	ψ	χ ψ
Omega	Ω	ω	ω

SI Units

TABLE B-1. Basic SI Units

Quantity	Unit	Symbol
time	second	s
mass	kilogram	kg
length	meter	m
electric current	ampere	A
thermodynamic temperature	kelvin	K
amount of substance	mole	mol
luminous intensity	candela	cd
plane angle	radian	rad
solid angle	steradian	sr

TABLE B-2 Special SI Derived Units

Quantity	Unit	Symbol	Formula
frequency (of a periodic phenomenon)	hertz	Hz	$1/s$
force	newton	N	$(kg \cdot m)/s^2$
pressure, stress	pascal	Pa	N/m^2
energy, work, quantity of heat	joule	J	$N \cdot m$
power, radiant flux	watt	W	J/s
quantity of electricity, electric charge	coulomb	C	$A \cdot s$
electric potential, potential difference, electromotive force	volt	V	W/A
capacitance	farad	F	C/V
electric resistance	ohm	Ω	V/A
conductance	siemens	S	A/V
magnetic flux	weber	Wb	$V \cdot s$
magnetic flux density	tesla	T	Wb/m^2
inductance	henry	H	Wb/A
luminous flux	lumen	lm	$cd \cdot sr$
illuminance	lux	lx	lm/m^2
activity (of radio-nuclides)	becquerel	Bq	$1/s$
absorbed dose	gray	Gy	J/kg

TABLE B-3. Some Common Derived Units of SI

Quantity	Unit	Symbol
acceleration	meter per second squared	m/s^2
angular acceleration	radian per second squared	rad/s^2
angular velocity	radian per second	rad/s
area	square meter	m^2
concentration (of amount of substance)	mole per cubic meter	mol/m^3
current density	ampere per square meter	A/m^2
density, mass	kilogram per cubic meter	kg/m^3
electric charge density	coulomb per cubic meter	C/m^3
electric field strength	volt per meter	V/m
electric flux density	coulomb per square meter	C/m^2
energy density	joule per cubic meter	J/m^3
entropy	joule per kelvin	J/K
heat capacity	joule per kelvin	J/K
heat flux density	watt per square meter	W/m^2
irradiance	watt per square meter	W/m^2
luminance	candela per square meter	cd/m^2
magnetic field strength	ampere per meter	A/m
molar energy	joule per mole	J/mol
molar entropy	joule per mole kelvin	$J/(mol \cdot K)$
molar heat capacity	joule per mole kelvin	$J/(mol \cdot K)$
moment of force	newton meter	$N \cdot m$
permeability	henry per meter	H/m
permittivity	farad per meter	F/m
radiance	watt per square meter-steradian	$W/(m^2 \cdot sr)$
radiant intensity	watt per steradian	W/sr
specific heat capacity	joule per kilogram-kelvin	$J/(kg \cdot K)$
specific energy	joule per kilogram	J/kg
specific entropy	joule per kilogram-kelvin	$J/(kg \cdot K)$
specific volume	cubic meter per kilogram	m^3/kg
surface tension	newton per meter	N/m
thermal conductivity	watt per meter-kelvin	$W/(m \cdot K)$
velocity	meter per second	m/s
viscosity, dynamic	pascal-second	$Pa \cdot s$
viscosity, kinematic	square meter per second	m^2/s
volume	cubic meter	m^3
wavenumber	one per meter	$1/m$

C

How to Make a Living as a Technical Writer

If you have a strong interest in science and technology and like to write, technical writing might be the profession for you.

Because many technical professionals write poorly, technical writers are needed in all areas of science, industry, and government. These are some of the jobs that technical writers handle:

Editing and proofreading copy

Ghostwriting trade journal articles

Working with engineers to help them improve their writing

Producing a wide variety of technical publications, including letters, memorandums, manuals, proposals, papers, reports, abstracts, product literature, advertisements, press releases, scripts, charts, and tables

Advising others in the organization about writing, graphics, printing, and binding methods

Providing authors with writing, editing, and research assistance

Preparing a writing style manual for the organization

Helping technical people with their speeches and presentations

Producing slide shows, films, and videotapes

You do not need a degree in science or engineering to write or edit technical publications. Although many full-time technical writers were scientists and engineers first, the majority came from the humanities, and the ranks are full of former English teachers, editors, journalists, and writers.

The would-be technical writer has three basic employment options: full-time, contract, and free-lance work.

Full-time technical writers hold staff positions with scientific and technical organizations. A technical writer at a large company might work in a group solely devoted to producing manuals, or proposals, or product literature. A technical editor at a trade journal works with contributing authors, preparing their manuscripts for publication. A small industrial manufacturer might hire one writer to handle all its technical communications.

The best place to find out about these jobs is the Help Wanted section of your local newspaper (look under Technical Writers, Writers, or Editors). As with any other professional position, you apply by sending your prospective employer a letter of application and a résumé. The one difference is that the résumés of technical writers stress descriptions of the types of publications they have handled, rather than a strict chronological listing of past employment by job title and company. A sample technical writing résumé appears on page 110.

Contract work is an attractive alternative to full-time employment. It offers the regularity of nine-to-five business hours without chaining the writer to one job with one organization.

When an organization needs extra technical personnel, it can contract their services through a ''temporary employment contractor.'' These employment contractors provide scientists, engineers, technicians, and technical writers on a temporary basis— for days, weeks, months, and sometimes years. The temporary employees work at the organization's place of business, but are paid by the employment contractor.

To get contract assignments, send several copies of your technical writing résumé to the employment contractor. You can find these contractors listed in the Yellow Pages under Employment Contractors, Temporary; Temporary Help; Technical Writing Services; or Editorial Services. The contractor will keep your résumé on file and will call you when they need to fill an assignment that matches your background and qualifications. Beginning technical

John Doe
100 Summertown Drive
Anyplace, USA
phone 000-0000

Technical Writer
Industrial Copywriter
Training Specialist

TECHNICAL WRITING EXPERIENCE

6/75-present   **ACE CHEMICAL COMPANY, Boonton, New Jersey**
Polymer mixing manual — produced installation, operation, and maintenance manual for a polymer mixing system used in injection molding operations. I had complete responsibility for the organization, data gathering, and writing of this manual through the printing stage, and I developed illustrations to explain theory of operation, wiring, and parts location. Customers found manual to be interesting, accurate, and easy to follow.
Other technical literature — although I am responsible for writing technical manuals from scratch, I also prepare technical papers, press releases, and product literature. For example, I edited and supervised the production and printing of a technical paper on the performance of motionless mixers.

6/72-5/75   **LIGHTNING ELECTRONICS CORPORATION, Paterson, New Jersey**
Shipboard radar manual — wrote a theory of operation manual on the W-120 shipboard fire control radar system. My responsibilities included setting up production schedule for typing and illustrations through the repro stage.
X-100 radar brochure — wrote and produced a glossy, four-color, 12-page sales brochure on the X-100 air traffic control radar system. Brochure describes capabilities, performance criteria, and operation of this airport surveillance radar.
Product information sheets — wrote product information sheets on the WX-200 modular shipboard fire control system, TCCS air traffic control communications system, X-100-AR air route surveillance radar, and other electronic systems.

ENGINEERING EDUCATION

9/68-5/72   B.S. in chemical engineering, University of Rochester, Rochester, NY.

RELATED ACTIVITIES

7/72-present   *Associate member*, American Institute of Chemical Engineers.
3/75-present   *Free-lance writer* — have published articles in SCIENCE BOOKS AND FILMS, Baltimore CITY PAPER, and THE ROCHESTER PATRIOT.

REFERENCES AND PORTFOLIO

Will be pleased to submit upon request

writers can expect to earn between $8 and $15 per hour; an experienced writer can make $20 to $25 per hour. (You can earn even more if you have experience in an unusual or a highly specialized technical field that is in great demand.)

Free-lancing offers even more freedom than contract work. As a free-lancer, you won't be locked into the corporate structure and the Monday-through-Friday workweek. Free-lancers can sleep until noon if they want to.

Every year, a quarter of a million people in the United States go into business for themselves. If you want to join them, you need to do a few things first:

1. *Put some money in the bank.* It takes times for any new business to show a profit. Before you leave the security of your job for the uncertainties of free-lance life, you should have enough money saved up to live for at least six months without any income.

2. *Decide exactly what you want to do.* What services will you offer your clients? writing? editing? graphics and printing? Are you strictly technical, or will you take assignments in other areas? You must decide what your business is—preferably before you start it.

3. *Promote yourself.* Now that you are on your own, you must go out and *get clients.* Free-lancers, like industrial manufacturers, need sales literature—a brochure or résumé that describes your business, the services you offer, your background and qualifications, and your fees. To get assignments, you could mail this brochure with a cover letter to organizations that could be potential clients. You might also promote yourself through publicity and advertising in technical magazines.

4. *Set your fees.* As a free-lance writer, you receive no bonuses, medical benefits, company insurance policies, sick days, or vacation. Therefore, your hourly rate must be higher than what you would get as a full-time employee. The going rate for free-lance technical writers ranges from $15 an hour for beginning technical editors to $50 an hour for experienced industrial advertising writers.

If you understand the basics of science and technology and can demonstrate an ability to write and think clearly, you should have no trouble making a good living as a technical writer, whether full-time, contract, or free-lance. Today the technical fields are booming, while the writing skills of college graduates are declining. Therefore, people who can write well on technical subjects are in demand. As a rule, technical writers earn slightly less than scientists and engineers, but more than writers and editors in nontechnical fields.

Where to Find More Information on Writing

In *Technical Writing: Structure, Standards, and Style,* we sacrificed some detail and scope to make the book easy to read and use. In selecting rules and topics, we used the "80-20 rule"—we present those 20 percent of the rules of usage that should answer 80 percent of the technical writer's questions.

When you need more detailed information, you can consult the books and publications listed below. We especially recommend the *U.S. Government Printing Office Style Manual* for its complete coverage of numbers, grammar, punctuation, abbreviation, and capitalization, and the ASTM *Standard for Metric Practice* for its thorough presentation of SI units.

American Society for Testing and Materials (ASTM): *Standard for Metric Practice,* Philadelphia, document E-380-79, 1980.

Angione, Howard (ed.): *The Associated Press Stylebook and Libel Manual,* Associated Press, New York, 1977.

Bernstein, Theodore M.: *The Careful Writer: A Guide to English Usage,* Atheneum, New York, 1967.

————: *Watch Your Language,* Channel Press, Manhasset, N.Y., 1968.

Berry, Thomas Elliot: *The Most Common Mistakes in English Usage,* McGraw-Hill, New York, 1971.

Flesch, Rudolph: *The Art of Readable Writing,* Harper & Row, New York, 1949.

Follett, Wilson: *Modern American Usage,* Hill & Wang, New York, 1966.

Miller, Bobby Ray (ed.): *The UPI Stylebook,* United Press International, New York, 1977.

Morrisey, George L.: *Effective Business and Technical Presentations,* 2d ed., Addison-Wesley, Reading, Mass., 1975.

Smith, Terry C.: *How to Write Better and Faster,* Thomas Y. Crowell, New York, 1965.

Strunk, William, Jr., and E. B. White: *The Elements of Style,* 2d ed., Macmillan, New York, 1972.

U.S. Government Printing Office (GPO): *U.S. Government Printing Office Style Manual,* Washington, 1973.